DISCARD

Watershed Management
Issues and Approaches

Watershed Management
Issues and Approaches

Timothy O. Randhir

Publishing

Published by IWA Publishing, Alliance House, 12 Caxton Street, London SW1H 0QS, UK

Telephone: +44 (0) 20 7654 5500; Fax: +44 (0) 20 7654 5555; Email: publications@iwap.co.uk
Web: www.iwapublishing.com

First published 2007
© 2007 IWA Publishing

Printed by Lightning Source

Disclaimer

British Library Cataloguing in Publication Data
A CIP catalogue record for this book is available from the British Library

Library of Congress Cataloging- in-Publication Data
A catalog record for this book is available from the Library of Congress

ISBN: 1843391090
ISBN13: 9781843391098

Contents

Preface

Any river is really the summation of the whole valley. To think of it as nothing but water is to ignore the greater part.

..Hal Borland: This Hill, This Valley

A watershed perspective looks at the whole landscape to address natural resource issues. This approach is becoming a common practice among communities and resource managers throughout the world. Environmental problems such as habitat loss, water contamination, dwindling freshwater supplies, desertification, urbanization, nonpoint source pollution, and ecosystem impairment are often a result of complex processes that require systems-based thinking, a central concept in watershed management. A watershed-based approach offers an excellent scope for the assessment and management of these problems. National and international agencies and researchers identify watersheds as providing a sound basis for developing an integrated, holistic, problem solving framework to restore and maintain the physical, chemical, and biological integrity of ecosystems, to protect human health, and to provide sustainable economic growth.

The last three decades saw tremendous progress in environmental awareness, scientific knowledge, and policy making. Natural resource scientists and policy makers are in general agreement that there is now a need for more integration of this kind of knowledge so that communities can better manage watershed resources. Watershed management is also increasingly being recognized by policy makers, resource managers, and communities as an effective way to achieve multiple goals and sustainability.

Managers involved in the sustainable management of watershed systems need to process vast scientific information to identify tools from multiple disciplines, to understand problems, and to identify appropriate solutions. In addition, the public involved in watershed decision-making needs to have complex information translated into simpler forms. The integration of a variety of information is central for quick and appropriate responses to problems in watersheds. The assimilation and use of watershed information are also important because natural resources and the environment are being impacted at a rapid and increasing pace, often accelerated by factors that include population growth and a higher demand on watershed systems for food, fiber, shelter, and services.

Restoration and maintenance of the physical, chemical, and biological integrity of watersheds require decision making that balances human and natural systems. The sustainable use of natural resources and services also requires achieving long term economic prosperity without impairing ecological integrity.

A balance between economic and environmental objectives and consideration of all interactions of the watershed system are important criteria in watershed management. This balance is necessary for countries at various stages of development. In spite of its importance and widespread applicability, there is a severe dearth of available, comprehensive information for the rapid assessment and solution of watershed problems.

The primary purpose of this book is to provide concise information on typical watershed problems by giving background information, methods to identify each problem, solution alternatives and suggestions for obtaining further information. Problem solving is a vital part of watershed management, wherein a particular issue is addressed within the scope of wider landscape and ecosystem connections, in the search for appropriate solutions. A problem solving approach will allow the rapid evaluation of these dimensions, sources of problems, their impacts, and appropriate solutions. Given the breadth of this subject, it is often difficult to find a single source of information for rapid assessment and management of natural resources. This approach will also be useful to local communities, researchers, policy-makers, resource managers, and educators. This book aims to provide the breadth of information on issues and concepts useful for watershed management.

This book is intended for natural resource professionals, students, and other nonprofessionals to learn about and use principles of watershed science in environmental problem solving. This book is also a good source for technical information on most watershed problems linked to natural resource conservation and planning. It can be used in teaching introductory courses in watershed management at the high school or undergraduate levels and also can serve as a reference for problem solving or inquiry-based teaching methods.

The problem-solving framework used here systematically focuses on each issue to identify and derive potential solutions within a watershed framework. The book is divided into six chapters, each covering a major area in watershed management. An introduction to basic concepts in watershed science is covered in the first chapter. This section provides information on hydrology, the water balance, watershed delineation, the water budget, assessment, planning and restoration. This chapter introduces watershed management principles that are used in the following chapters. In the second chapter, various land use issues and their implications for the watershed system are covered. In this chapter each problem associated with land use is described, followed by an explanation of observable symptoms and their consequences. This is followed by a discussion of available structural and nonstructural solutions to manage the problem. The third chapter focuses on problems associated with inland water bodies, especially rivers, lakes, ponds, wetlands and vernal pools. The fourth chapter covers various issues facing coastal watersheds such as sediment loading, shoreline stabilization, eutrophication, development, and shellfish and beach contamination. The fifth chapter discusses aspects of ecosystem health and biodiversity, landscape ecology, stream and river corridors, open space and species richness. The last chapter addresses watershed problems and issues associated with water quantity and quality. To allow consistency and quick reference, a common format is used in each topic: (i) Description; (ii) Problems and Impacts; (iii) Solutions: a. Structural; b. Nonstructural; and (iv) Suggested Reading. My intent is to cover the breadth of information necessary for watershed management. While each topic is discussed in a summarized format, readers needing more information on a topic can benefit from the case studies, and suggested readings.

This book is a result of several years of experience in working with watershed management issues. I am thankful to my graduate students Eric Marshall, Debbie Shriver, Greg Writtenaur, Olga Tsvetkova, Paul Ekness, Elinor Keeler, Kerri Davis, Sekar, Sarah Low, and Michelle Matteo who assisted me with the compilation and editing of material on this vast subject. I would like to thank Debbie Shriver for meticulous proof reading and formatting. I also wish to acknowledge the support and comments provided by Robert O'Connor of the Massachusetts Executive Office of Environmental Affairs.

I would like to thank my wife Reena and my children Priyanka, Ashwin, Richita, and Nivedita for being a part of this project through their encouragement, patience, and support. This book would not have been possible without the guidance and grace of God.

Timothy O. Randhir

Dedicated to my parents and grandparents

1

Introduction: Watershed Basics

1.1 WATERSHED: DEFINITION and DELINEATION

1.1.1 What are watersheds?

Watersheds represent a natural way of dividing the landscape for management and planning purposes. Watersheds are generally referred to as drainage areas, and the boundaries are delineated using changes in elevation of the landscape. Most land is a part of a watershed. The mountains, valleys, plains, and streams are natural features of a landscape that direct and distribute water and are components of the watershed ecosystem. Watershed boundaries do not generally coincide with administrative, property, and political boundaries. However, watershed boundaries should be considered in community based decision making (Figure 1.1). Rivers, lakes, estuaries, wetlands, streams and the coastal portions of oceans are water bodies that are of interest in natural resource management. The status of these water bodies is closely related to upstream land use within the watershed. The actions of people living within a watershed thus affect the quality and quantity of waters that drain from it, thereby having an impact on natural systems and regional economics.

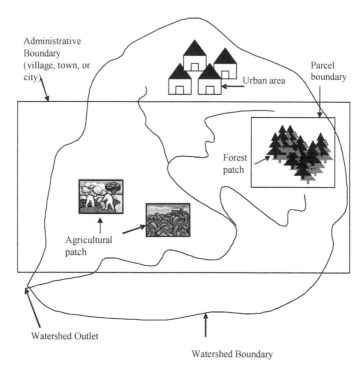

Figure 1.1 Watershed and community boundaries

John Wesley Powell, a scientist and geographer, defined the watershed best as "that area of land, a bounded hydrologic system, within which all living things are inextricably linked by their common water course and where, as humans settled, simple logic demanded that they become part of a community." (Powell 1890).

A watershed is commonly defined as the area of land that collects precipitation in the form of rain and snow and discharges or allows it to seep into a marsh, stream, river, lake or groundwater (Figure 1.2). A watershed is defined by the area that drains to an outlet point, often starting at a ridge line in the headwater areas, and including the land that drains into streams and tributaries, to higher order rivers, and into downstream water bodies.

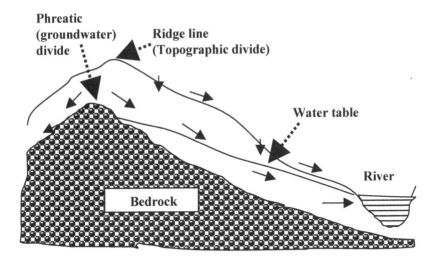

Figure 1.2 Water flow in a watershed

The watershed is increasingly gaining recognition as a hydrologic and ecological unit of natural resource planning and management. This is because of the watershed's relatively stable boundaries, the increasing use of a systems approach in natural resource planning and management, grassroots support from communities, and because a watershed approach taps into peoples' awareness of landscape features. The application of a watershed-based strategy to land management problems can aid in the understanding of natural and human-induced stresses of natural resources. The watershed can also provide a useful basis for comprehensive assessment and planning for the sustainable use of natural resources. A systems approach in a watershed context can be highly effective in natural resource planning and management compared to methods that focus on single components of a watershed system, such as a stream, lake, or a single discharge point.

1.1.2 Watershed Delineation

Delineation is the process of identifying the boundaries of the watershed and is an important step in watershed management. Watershed boundaries are often identified on elevation maps by starting at the outlet point and marking ascending contours up to the ridge line. Black (1996) suggests three rules for delineation of watersheds using contour maps: 1) Water tends to flow perpendicularly across contour lines; 2) Ridges are indicated by contour V's pointing downhill; 3) Drainages are indicated by contour V's pointing upstream.

Geographic Information Systems (GIS) can be used to derive accurate watershed boundaries.

A common procedure for the rapid delineation of a watershed is to use a contour map (Figure 1.3). Start by marking the outlet or downstream point of the water body. All high points are then identified along both sides of the watercourse. Often this mapping can be done by moving upstream toward the higher points in the watershed. The watershed boundary is identified by progressively connecting these high points, starting on one side of the outlet and ending on the other, making sure that the line is perpendicular to the contours. The watershed boundary starts at the outlet and connects the ridge lines on both sides of the stream and then continues back to the outlet. The accuracy of the delineated boundaries can be checked by paying close attention to the elevation values of the contours.

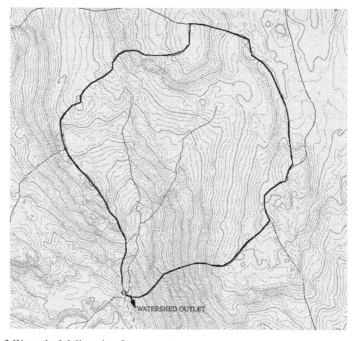

Figure 1.3 Watershed delineation from a contour map

Watersheds are ecological units that are nested hierarchically; that is, watersheds can be subdivided (aggregated) to lower (higher) sized watersheds. At a global scale, there are 114 major watersheds (World Resources Institute 2003) draining major rivers of the world. These watersheds can be further delineated into sub-watersheds based on their major tributaries, and further subdivided to the first order streams that are in the headwaters of a watershed.

Thus watersheds are nested at various scales. At a national scale, such hierarchical classification of the watersheds is common. For example, the United States Geological Survey (USGS) system divides the U.S. (Figure 1.4) into 21 regions, 222 sub-regions, 352 accounting units, and 2,262 cataloguing units. A hierarchical hydrologic unit code (HUC) consisting of 2 digits for each level in the hydrologic unit system is used to identify any hydrologic area. The 6-digit accounting units and the 8-digit cataloguing units generally refer to a basin and subbasin. This system is defined in the U.S. Federal Information Processing Standard (FIPS) and serves as the backbone for hydrologic delineation. Each watershed in the U.S. Environmental Protection Agency's (U.S. EPA's) Surf Your Watershed program is defined by the 8-digit cataloguing unit.

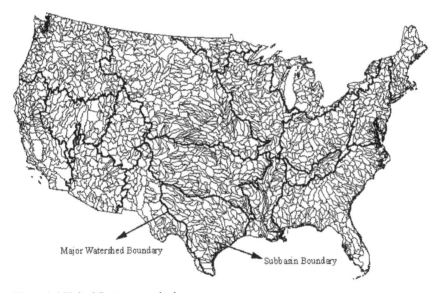

Figure 1.4 United States watersheds

1.2 WATERSHED APPROACH

Why Watersheds?

A watershed approach can be defined as an integrated framework for environmental and natural resource management that coordinates public and private efforts for decision making and planning and which considers the hydrologic cycle, ecosystem dynamics, and socioeconomic and political characteristics. There are several advantages in using a watershed approach for planning and management:

- *Natural management units* Watersheds are based on natural hydrology and have relatively fixed boundaries. They are often recognized as practical units for understanding the interconnections among ecology, geography, geology, and cultural features that affect land and water. While defined through hydrological means, watersheds are increasingly being used as ecological and regional units to manage resources as a part of a system.

- *Better results* We can expect better results when the resource interactions and usage are understood well. Since watersheds are based on natural hydrological and ecological processes, resource managers can better understand and evaluate underlying problems and conditions and develop comprehensive solutions. Such results are often better for achieving sustainable solutions than fragmented and end-of-pipe methods.

- *Ecosystem-based and comprehensive* A watershed approach uses all components of the system and their interactions in evaluating and assessing natural resource problems. These components also include economic and human interactions and are thus more comprehensive than a non-system approach. The watershed-based system boundary allows a better understanding of the interaction among the abiotic (soil, water, and air), biotic (plants, animals, and human), and socioeconomic (markets, technology, and other institutions) elements.

- *Economic efficiency* A watershed approach encourages cooperation and collaboration among government, business and citizens, and thus aims at long term, sustainable solutions. These associations can result in substantial savings in time and resources. In addition, a watershed approach identifies potential impacts throughout the system and can be used to avoid mistakes. Watershed-scale markets can also be used to achieve cost-effective reductions in pollutants.

- *Public support* Community awareness and participation are emphasized in a watershed approach, which develop a sense of community to achieve environmental goals. When people are involved in decision making and management, the likelihood of conflicts is reduced. A good watershed plan increases the commitment of the public in protecting and restoring their natural resources.

- *Grassroots planning* A watershed approach encourages "bottom-up" planning with the recognition of all stakeholders and their interests at the grassroots level. This recognition encourages public participation in decision making. A participatory process should also incorporate maximum information on the societal issues of the region to avoid bias and the dominance of private interests.

- *Watershed units are readily recognized* by communities, and therefore people can relate to their landscapes much more easily. A sense of belonging to a landscape can encourage community-based planning and management and voluntary efforts.
- *Encourages interdisciplinary and interagency cooperation* A watershed approach can encourage researchers and agencies to cooperate by providing a common framework for their efforts. Most watershed research is multidisciplinary, and this leads to a high degree of inter-agency cooperation in watershed-based projects.
- *Administrative streamlining* Watershed-based projects allow better planning and streamlining of resources, a reduction in reporting needs, and targeted financial assistance.
- *Demonstration of successes* There are numerous examples of success in watershed-based planning in various parts of the U.S. and the world. The projects recognize the uniqueness and advantages of this approach. Comprehensive assessment and management encourage conservation and efficient resource use, and when well planned, can be used to achieve sustainable resource use.

1.3 HYDROLOGIC CYCLE

Water is the source of all life and is often distributed unevenly, which makes it important to watershed management. The circulation of water is called the *hydrologic cycle*, and this is the fundamental framework of watershed management (Figure 1.5).

The continuous transfer of water from the atmosphere to the earth's surface, into the groundwater and surface water systems, and its return to the atmosphere is a process collectively called the hydrologic cycle. This cycle occurs in the biosphere and collects, purifies, and distributes the earth's water supply. Change or disruption of the cycle often results in changes in the status of landscapes and ecosystems. The hydrologic cycle can be divided into two portions: the atmospheric branch in which movement of water is in a gaseous phase, and the terrestrial branch, where water flows mainly as a liquid.

Solar energy drives the hydrologic cycle. When heated by the sun, the surface molecules of water are energized and start the process of evaporation, rising as water vapor into the atmosphere. Approximately 80 percent of the evaporation is from oceans. The direct evaporation process accounts for 84 percent of atmospheric water vapor. Water vapor is also exchanged from plant leaves through a process called transpiration, which occurs through the stomatal openings (air sacs) on the surface of plant leaves. The water vapor then collects into the clouds. Surface circulation and jet streams can move these clouds over long distances on the earth's surface.

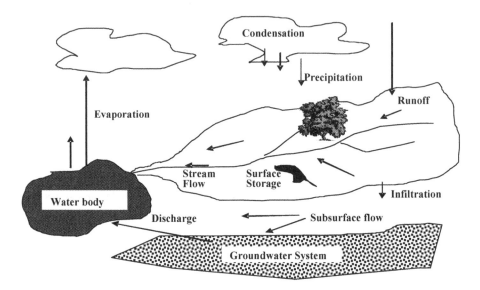

Figure 1.5 Hydrologic cycle

When the atmosphere is saturated with water vapor, water changes from a gaseous to a liquid phase. As water vapour moves across the earth, it rises and cools until it condenses around atmospheric dust to form liquid water droplets, which may fall as precipitation (rain, hail, snow, sleet, freezing rain). Processes that trigger precipitation include orographic (influence of mountain ranges), frontal systems (interface between high and low pressure systems), and convective processes (heating of the earth's surface). As precipitation reaches the earth's surface and flows down through a watershed, it moves down gradient following the quickest paths. The flow travels through the drainage system formed by the movement of water in the watershed, which starts in small streams, drains into larger rivers, and eventually discharges into the ocean.

Depending on the land cover in a watershed, some water infiltrates into the ground through the soil and reaches groundwater. The groundwater consists of water that lies below the land surface in openings between soil particles, fractures, and porous rock strata. Water-bearing porous rock and soil strata are called aquifers. Groundwater can be connected through channels that allow water to flow slowly through layers of sand, gravel, and porous rock. Groundwater forms the major source of freshwater in the world and is often tapped by pumping it from wells for human uses.

Both overland flow and groundwater eventually discharge into streams that empty into the ocean or other large water bodies. Plants can take up the water or it evaporates from the surface, and thus the cycle continues.

Case Study

Watershed Name: Lake Titicaca Basin (TDPS System)
Location: Covers parts of Peru, Bolivia, and Chile, South America
Major Problem: Structural poverty and climate variability affecting tropical glaciers
Approaches: Address water problems within the greater social framework. The plan emphasizes better management of land, water, and gas resources.
Information:
http://www.unesco.org/water/wwap/wwdr2/case_studies/pdf/lake_titicaca.pdf

---

1.4 WATERSHED COMPONENTS

The watershed is an ecosystem where biotic (plants, animals), abiotic (soils, water, air), and socioeconomic components interact. Successful watershed planning must consider all of the socioeconomic, biophysical, ecological, regional, and political elements.

The physical components of a watershed include its biophysical parts and land uses through which water drains into a riverine system. The watershed begins at the headwaters, which are usually at the highest elevation in the watershed, such as the top of a mountain range or hill. The highest elevation areas (ridge lines) mark the physical boundaries between different watersheds. Water flowing down the opposite slopes of ridge lines usually feeds different drainage basins. As water flows down to lower elevations, it gains volume and velocity and erodes soil to form well-defined creeks, brooks, and rivers. The network of the drainage in a watershed varies in structure depending on the geological history and composition of the watershed. Since headwaters are usually fed by precipitation, the headwater flows often have good water quality compared to other places in a watershed with similar land uses. Because headwaters have very small streams (first order streams), they are also the most vulnerable to human disturbances. These small streams often respond rapidly to changes in land use and disturbance conditions.

Riverine systems consist of a network of water bodies (rivers, lakes, ponds, and streams) that receive and accumulate flows from smaller streams in a watershed. The riverine system is the main flow of water exiting the watershed into a larger river system or into the ocean.

Wetlands are defined as areas saturated with water that have unique plants adapted to waterlogged conditions and which have hydric soils (saturated and anaerobic). Wetlands are important components of a watershed ecosystem that perform unique hydrologic, biological, and economic functions.

The *floodplain* of a river system is the flat land area of the valley floor that is adjacent to a river and which is susceptible to being inundated by water. The floodplain often serves as a natural flood and erosion control structure during regular flooding. Riparian vegetation that occurs at the water's edge filters out sediments and other contaminants from runoff before they enter the river. The dense vegetation also provides a canopy over the river, lowering the water temperature in shallow areas of the stream. In addition, during flooding, the dense vegetation in a riparian zone can regulate flows by slowing the velocity, temporarily retaining, and slowly releasing water back into the river system.

Benchmark watersheds are areas (subwatersheds or patches) within the watershed that are relatively less disturbed by human activities (Figure 1.6). These areas can form a reference point for the potentially attainable habitat status, especially during an assessment of habitat conditions elsewhere in the watershed. Benchmark watersheds also serve as a performance target for restoring disturbed watersheds.

Biotic refuges are pockets that maintain the biodiversity of the watershed because the habitat is still undisturbed and relatively healthy. Many endangered species depend on these biotic refuges when their habitat is almost completely destroyed elsewhere.

Biological hot spots are intact patches of riverine habitat that provide critical functions to the ecosystem. They contain a rich, but threatened diversity of plant and animal life. Urbanized areas have high levels of human-induced degradation in these watershed ecosystems. Benchmark watersheds could be used as references in the restoration of biological hot spots (Doppelt et al. 1993).

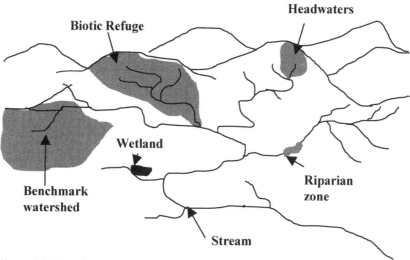

Figure 1.6 Watershed components

Managed rural components include *agricultural lands* and *managed forests* whose operations influence watershed soils and vegetation.

The socioeconomic components of the watershed ecosystem include *human components* that are represented by population density and flux, resource values, markets, belief systems, employment patterns, and products and services. Community interactions, institutions, organizations, rules and regulations, and societal values are all important parts of a watershed system.

1.5 WATER BUDGET

An important concept in watershed management is the watershed budget that represents the hydrologic balance of water in the biosphere. This concept is derived from the law of conservation of mass. The status of the watershed budget is critical in diagnosing and assessing a problem facing the watershed. Evaluating the water budget is a method to assess the nature of the hydrologic cycle and to quantify the impact of a particular disturbance in a watershed. Evaporation, precipitation, infiltration, storage and runoff are major components of the water budget.

The water budget equation is expressed with the formula:

I – O = ΔS, where, I is inflow, 0 is outflow, and **ΔS** is change in storage.

In normal circumstances, the inflow of water to a system equals the outflow, and there is no change in water storage. During droughts and floods, this balance can be upset and the amount of water in storage can change. The water budget can be used to quantify water resources of a region and to determine the potential vulnerability of the hydrologic cycle to droughts and floods and to assess the implications for land use, ecosystems, and regional economics.

Case Study

Watershed Name: Tuul River Basin
Location: Mongolia, Asia
Major Problem: Poverty, access to safe water, and sanitation facilities
Approaches: Water conservation as reflected in National Water Programme. Decentralization of water pricing has promoted economic growth by providing low cost water to industry.
Information:
http://www.unesco.org/water/wwap/wwdr2/case_studies/pdf/mongolia.pdf

---

1.6 WATERSHED ASSESSMENT

Watershed assessment is used to evaluate the state of the system and is an important part of watershed management. An assessment provides critical information that is vital to quantify the impacts of land use, to evaluate progress, and to develop efficient watershed plans. Watershed assessment can be used to determine various changes in water resources, ecosystem health and productivity, and the economics associated with land management practices. For instance, watershed assessments can be used for the evaluation of proposed urban development, expected economic growth, or future land use on peak flows, suspended sediment, bed load, and stream channel stability within the watershed. Watershed assessment requires the compilation and classification of data on abiotic, biotic, and socioeconomic components of the watershed.

Compiling a watershed inventory requires both historic and current data on various components of the watershed. Often a classification based on abiotic (soils, water, air), biotic (plants and animals), and socioeconomic (human activities, economics) factors can be used to categorize and assess the status of the watershed. Analysis of the data and decision making often depend on information obtained from watershed assessment. Modeling is becoming an integral part of watershed management and is being used by professionals and agencies throughout the world. Developing a watershed model using geographic information systems (GIS) and simulation modeling is an effective approach for determining alternative solutions and decision making in watersheds. Information obtained from remote sensing is also playing a major role in providing watershed information. A limitation, however, is that GIS data are not always available in many parts of the world, and compiling this information can be expensive and time consuming. Community mapping, where volunteers develop hand drawn watershed maps, can be used in areas with poor GIS data. Where GIS is available, data models have been used to assess levels of imperviousness, erosion, nonpoint source pollution, runoff, ecosystem changes, and economic impacts. Models of a benchmark watershed can be used as a baseline reference point for simulation.

1.7 WATERSHED PLANNING

Watershed management is often complex and challenging because of the amount of information that must be processed and the many tradeoffs to be considered. For example, land use and water quality impacts should be assessed as a system within the framework of the socioeconomic and policy environments. Watershed management practices thus have far reaching impacts on all parts of the watershed system. Watershed planning necessitates a careful analysis and integration of economic and ecological implications in order to achieve a sustainable outcome.

A typical watershed planning process consists of five major steps to assess and develop a comprehensive watershed strategy. A protocol (Randhir and Genge 2005) for developing such a plan is presented in Figure 1.7. It starts with a complete inventory of resources, potential sources of contamination, the nature of pollution transfers, and development of a community profile. For various pollutant sources, structural and nonstructural solutions are listed along with their benefits and costs of achieving improvements in water resources. In the next step, the practical implications and feasibility of each solution, along with cost considerations, are evaluated to identify a set of best strategies to achieve water quality improvements. Finally, the watershed plan is completed by including feedback and contingency alternatives. After an evaluation of all the information, the next stage develops an institutional mechanism that can implement this plan. A plan needs to engage communities and stakeholders in the watershed and be adaptive in nature. An adaptive process incorporates new information, changes in the political and economic situation, and it addresses shortcomings of a plan.

The objectives of the watershed plan should be clearly defined based on the problems and opportunities available. The assessment and prioritization of watershed problems is often a difficult task, requiring extensive monitoring and an evaluation of past activities. An element of the planning process is the evaluation of external constraints, which limit the scope of feasible options. Constraints may include social, political and cultural limitations, and the scarcity of economic resources.

Prioritizing watershed problems and decision making require an evaluation of many alternative solutions and ranking them according to feasibility, effectiveness, and cost. A ranking using spreadsheets or mathematical programming can be used to identify optimal strategies. Strategies to solve watershed problems should be adaptive and incorporate community views. Adaptive management can be accomplished through continuous monitoring of watershed performance (such as water quality and biodiversity), ongoing discussions with stakeholders and the public, updating information, and incorporating better decision tools.

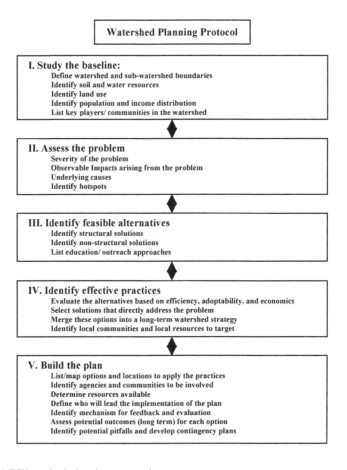

Figure 1.7 Watershed planning protocol

Watershed planning is an evolving discipline, and spatial information for comprehensive management is still scarce. The emergence of computer models and use of the Internet have enabled better decision support for communities and watershed managers. Data and information on the effectiveness of different approaches can be more easily shared and disseminated. A participatory approach to planning should involve all stakeholders throughout the planning cycle. This process can encourage voluntary efforts, reduce conflicts, and empower all players in the watershed.

Watershed managers can use the "with and without" concept in an evaluation of alternative projects. In this approach, changes in watershed health and productivity over time are assessed for two scenarios: one in which the particular management project is undertaken, and other one in which it is not. This method requires an in-depth understanding of economic and ecosystem functions, extensive data collection, and analysis of alternatives.

Case Study

Watershed Name: South Fork Tolt Watershed
Location: Washington, U.S.A.
Problem: Water quality
Approaches: Watershed education program in city's schools.
Information:
http://www.epa.gov/safewater

1.8 WATERSHED AS A MANAGEMENT UNIT

Watersheds are defined by natural, hydrologic boundaries and allow a systems-based approach. An important aspect of a system is its boundary, and watersheds have a relatively stable boundary compared to other regional units such as biomes and patches. (In unusual circumstances, the watershed boundary can change through landslides, erosion, natural disasters, and human activities.) Watersheds are also hierarchical in structure and can be aggregated at larger scales or divided into smaller units at finer scales. Thus, watersheds are increasingly gaining acceptance as units of ecosystem management. Within the watershed a management focus on ecosystem processes will aim for sustainable land use and resource yield compatible with the watershed's aesthetic and ecological conditions. Sustainable development that "meets the needs of the present generation without compromising the ability of future generations to met their needs" (WCED 1987) is now an acknowledged goal of watershed management. An ecosystem approach to watershed management is comprehensive and is based on the biological and physical resources of the watershed and considers the economic well being of human communities over multiple generations to achieve sustainability.

Because watershed management encompasses all processes within the system, the watershed is therefore a logical scale for managing water resources. Using an ecosystem approach, the whole system becomes the focal point, and managers are able to gain a more complete understanding of the interactions and overall impacts of the stresses. The watershed ecosystem approach manages resources through an integrated analysis of the links between the processes and patterns of watershed use.

The need for such comprehensive management of resources through an integrated watershed approach is emphasized in Agenda 21 (United Nations 1992). Agenda 21 is a comprehensive plan of action to be taken globally, nationally, and locally by organizations to manage human impacts on the environment (United Nations 1992). A watershed approach reflects the goals of Agenda 21, which addresses a variety of issues. Agenda 21 has global support and was strongly reaffirmed at the World Summit on Sustainable Development held in Johannesburg, South Africa in 2002.

1.9 WATERSHED RESTORATION

Most watersheds are impaired in some way or other by human activities. From minor forest management to intense urban activities, watersheds face varying levels of impacts. Restoration is becoming an important area of watershed management, where a watershed is restored to a certain, desired level (often decided by the manager or communities).

Watershed restoration can be defined as improvements in structure (biophysical components) and function (processes) of the watershed towards an ideal level. The process is presented in Figure 1.8.

The current status of the watershed system is often defined by the structure and function it maintains. Watershed restoration requires an approach that balances both structure and function (as shown in the figure) along the line that tends toward the historic conditions of the watershed. Restoration that moves the status of the watershed to the lower-right portion of the quadrant is inefficient because the restoration is higher in structure and lower in function. This imbalance between structure and function could lead to the failure of the restoration investments. An example of this is installing plant species in physical conditions that are similar to historic conditions without concern for current ecosystem processes and overall sustainability of the system. Restoration activities that move the watershed toward the upper-left area of the quadrant lead to higher function with poor structural characteristics, again leading to inefficiency. An example of this is to mimic watershed processes using artificial means without developing a sustainable structure that moves closer to historic trends. For example, developing a flood mitigation function through a flood diversion structure can be used to rapidly drain water out of a watershed, but may not be sustainable in the long run because of loss in riparian ecosystems and groundwater recharge.

Often, restoring to historic conditions is not practical and could involve high costs of restoration, especially in urbanized watersheds. In these situations, one approach is to focus on small changes (short run strategies) that could lead to a long-run strategy that moves closer to historic conditions. One problem is the identification of a historic or ideal baseline to use in restoration planning. This restoration goal should be decided through consultation with and education of the stakeholders in the watershed.

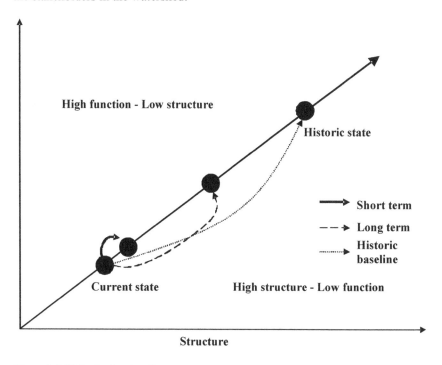

Figure 1.8 Watershed restoration process

The principle of "protect the best, and restore the rest," can be used as a strategy for effective watershed restoration efforts. This method focuses on preventive and restorative strategies. Some examples of preventive strategies include protecting headwaters and riparian areas from urbanization and disturbances. Identifying problems and prioritizing them for restoration can be very difficult in large watersheds. The cumulative and system-wide effects of a problem need to be assessed for prioritization. Water sampling can give clues to the kind of pollution and its pathway of contamination in the watershed.

Point source pollution, such as that from paper mills, cotton bleachers, and wastewater treatment plants, is easily detected and monitored. This type of pollution is often regulated by legislation on pollution levels, thereby forcing the industry to commit to the adoption of improved technologies. Pollution problems also occur from nonpoint sources, which are diffuse in nature and difficult to manage. Nonpoint source pollution is the result of runoff waters from such activities as agriculture, forestry, and urban development. Proper land use practices and best management practices (BMPs) can be used to minimize this type of pollution, but are not often adopted because they increase costs to adopt conservation practices by private individuals. Getting land users to follow improved land use practices requires incentives and information about nonpoint source pollution.

Case Study

Watershed Name: Basque Community
Location: Northwest Spain, Europe
Major Problem: Flooding
Approaches: Integrated flood prevention plan. A hydro-meteorological monitoring network has been established to assess aquatic ecosystems and regional water bodies.
Information:
http://www.unesco.org/water/wwap/wwdr2/case_studies/pdf/basque_country.pdf

---

1.10 ORGANIZATIONAL ISSUES – A SPECIFIC CASE

Organizational issues in watersheds revolve around management of limited resources to achieve the goals of an organization. Often organizations in watersheds face numerous challenges that include a lack of coordination among agencies, poor public support and participation, deficient information about the components of the watershed, budgetary constraints, insufficient scientific data and guidance, a lack of comprehensive objectives, and the absence of a systems-approach to planning. Given that watershed organizations are as diverse as watersheds themselves, this topic is discussed using a specific case in the U.S. The Massachusetts Watershed Initiative was a state program to organize activities at a watershed scale. The Commonwealth of Massachusetts's Executive Office of Environmental Affairs fostered a watershed-based partnership among local communities, governmental agencies, and volunteer organizations in Massachusetts, U.S.A. The program recognized the need for integrated, cooperative approaches in which participation and interaction among community, government and business are essential. Massachusetts was divided into 27 watersheds (Figure 1.9), and each watershed had a team represented by local, state, and federal agencies as well as community and business representatives. Each team was headed by a full time Watershed Team Leader. Each Watershed Team worked on a five-year cycle, focusing on one goal each year. The five focus goals of the watershed cycle were: Outreach, Research, Assessment, Planning/Implementation, and Evaluation. Each of the 27 watersheds started at different stages of the cycle in order to share the financial resources available in any year.

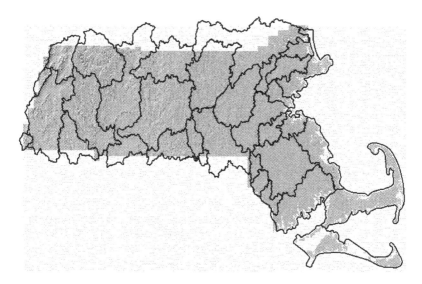

Figure 1.9 Watersheds in Massachusetts, U.S.A.

The success of the Watershed Initiative relied entirely on the cooperation and participation of many different stakeholders. Cost sharing among agencies had also become an important tool to complete important, but expensive projects. A comprehensive five-year watershed plan was developed during the assessment phase of a watershed. This plan helped to guide state and federal grants and loans to the most important environmental problems affecting the communities. After five years of successful implementation and extensive improvements to watersheds throughout the state, the Massachusetts Watershed Initiative lost its funding due to state budget cuts and the organization was dissolved. Some of the efforts continue to function through volunteer organizations (watershed coalitions) in some of the watersheds.

1.11 THE U.S. CLEAN WATER ACT

A national-level organization of environmental efforts is critical to provide guidance and oversight on the state of watersheds. One such effort in the U.S. is the Clean Water Act of 1972. The U.S. Congress enacted the law "to restore and maintain the chemical, physical, and biological integrity of the nation's waters." The national goal is for all of the nation's waters to be safe for fishing and swimming. While substantial progress has been made in improving water quality in most watersheds in the U.S., the goal has not been achieved in many water bodies. To achieve this ambitious goal throughout the nation, a variety of programs have been designed and implemented to address many sources of pollution. Point sources, which are concentrated discharge points from large, industrial sources and sewage treatment plants, are now regulated well, while nonpoint sources continue to be a major challenge. Nonpoint source pollutants include runoff from storms, which contains pesticides, manure, fertilizers, oil and gas, and other types of contaminants from farms, ranches, streets and parking lots, suburban lawns, and other large areas.

The Clean Water Act has established federal and state partnerships to control the discharge of pollutants from large point sources through a four-step process:

(1) The U.S. Environmental Protection Agency develops national guidelines for the control of industrial pollution discharges. Regulations are often based on the "Best Available Technology" (BAT) that is economically achievable in all industrial categories.

(2) Individual States establish the water quality standards necessary to make the water bodies in their state clean enough for their designated uses such as fishing, swimming, boating, wildlife habitat, agriculture, and industry.

(3) Sewage treatment plants must meet basic levels of treatment that transform disease-causing organisms into harmless matter. The federal government has provided billions of dollars in grants and loans to state and local governments to construct such plants.

(4) All industrial and sewage treatment plants must obtain permits that specify the type and amount of pollutants they may discharge. These permits specify industry-wide technology standards, state water quality standards, and sewage treatment standards that apply to each source. Discharge permits are reviewed and renewed every five years to account for improvements in technology.

Nonpoint pollution from agriculture, forestry, mining and other land uses is a large source of water pollution. States guide landowners in planning and utilizing cost-effective, best management practices at the earliest practical date.

Urban stormwater is another source of water pollution. Cities with a population of 100,000 and heavy industries must have stormwater discharge permits that establish minimum requirements for preventing stormwater pollution. These permits require dischargers to develop pollution control plans that meet state water quality standards. Dischargers are held accountable for complying with their plans. The regulation has been extended to smaller towns, cities and operations in the second phase of the stormwater program.

1.12 TOTAL MAXIMUM DAILY LOAD

The U.S. Clean Water Act (section 303) also establishes the water quality standards and the TMDL (Total Maximum Daily Load) programs. The TMDL is an estimation of the maximum amount of a pollutant that a water body can receive and still meet water quality standards, and an allocation of that amount to the pollutant's sources. It is defined as the sum of the allowable loads of a pollutant from all contributing point and nonpoint sources. The calculation includes a margin of safety to ensure that the water body can be used for the purposes designated by the State, and it accounts for seasonal variation in water quality.

Water quality standards are set by states, territories, and tribes that identify uses for each water body (for example, drinking water supply, contact recreation such as swimming, and aquatic life support for fishing) and the level of water quality necessary to support that use.

1.13 PATHWAY ANALYSIS FOR WATERSHED INSTITUTIONAL DESIGN

Water and air pollution are major problems facing watersheds with cumulative effects that influence downstream locations such as coastal systems. One method to address such problems is to identify the nature of technological externalities (the physical influence of a watershed activity such as land use on other stakeholders) that impact water resources through hydrologic, ecological, and economic processes. Examples of negative externalities in a watershed include point sources of pollution which affect downstream water quality, oil spills by an industry which impact aquatic ecosystems, air pollution activities that create increased acidity of precipitation received in other areas, farm pollution from fertilizer applications which create a eutrophication problem in downstream water bodies.

A pathway analysis (Randhir and Genge 2005) systematically identifies paths of pollutant transfer and fate or ecosystem impacts from a source. Steps to apply a pathway analysis are:

- *Identify sources of externalities (locations)*
  A map or GIS can be used to identify potential sources of problems. Historic and emerging sources should be included in this mapping to identify emerging and past sources.
- *Identify polluting agents (decision makers)*
  A complete inventory of communities and property owners who are the sources of the externality must be assessed. For this, a survey and census data can be used. Another option is to use land use information to identify potential decision makers involved in the externality.
- *Track pathways of the externality (Contaminant or ecosystem impact)*
  This is an important step in the process. Elevation maps and landscape ecological assessments can be used to quantify the pathways of these sources. For example, contaminant pathways flow downstream and can be mapped using hydrologic flows and elevation modelling. Modelling can be an effective and accurate tool to quantify the nature of the pathways.
- *Assess effects of the externality*
  The impacts should be comprehensively assessed along each pathway. This assessment ought to include ecosystem and economic impacts of each source. While this process creates extensive information, it is useful to assess cumulative and offsite impacts of an action.
- *Identify all affected agents*
  Agents that are affected can be tracked along the pathway. An account of the nature of impacts is critical in addition to the pathway distance from the source.
- *Identify appropriate technologies to internalize the externality*
  Internalizing the problem at the source level is an effective solution to the problem. This can be achieved through implementation of management practices that minimize external impacts. Another option is to install practices along the pathway to mitigate the external impacts.
- *Build a cooperative mechanism for addressing the externality*
  Develop methods for joint efforts to address the externality by sources and affected agents. These efforts could include compensation for management practices, regulation, incentives through cost sharing and tax breaks, etc., that encourage sources to internalize the externalities. The pathway analysis requires the identification of alternative institutional arrangements to mitigate a specific externality in the watershed. The criteria that will be used in the development of alternative mechanisms require the appropriate combination of incentives and information access for voluntary efforts. This can be achieved through collective action at the watershed scale.

SUGGESTED READING

1.1 – 1.2 Watershed Definition, Delineation and Approaches

Black, P. (1996) *Watershed Hydrology*, Ann Arbor Press, Chelsea, Michigan.

Natural Resources Conservation Service (NRCS) (2006) *How to Read a Topographic Map and Delineate a Watershed*. U.S. Department of Agriculture, NRCS. URL: http://www.nh.nrcs.usda.gov/technical/Publications/Topowatershed.pdf.

Powell, J. W. (1890) Institutions for arid lands. *The Century*, **40**, 111–116.

United Nations (UN) (1992) *Agenda 21: Chapter 18, Protection of the quality and supply of freshwater resources: Application of integrated approaches to the development, management and use of water resources.* UN Department of Economic and Social Affairs, Division for Sustainable Development. URL: http://www.un.org/esa/sustdev/documents/agenda21/english/agenda21chapter18.htm.

U.S. EPA (2006g accessed) *Surf Your Watershed*. U.S. Environmental Protection Agency. URL: http://www.epa.gov/surf/.

U.S. EPA (2006i accessed) *Watershed Academy*. U.S. Environmental Protection Agency. URL: http://www.epa.gov/watertrain/whywatersheds.html.

U.S. EPA (2006k accessed) *Watershed Approach Framework*. U.S. Environmental Protection Agency. URL: http://www.epa.gov/owow/watershed/framework.html.

U.S. EPA (2006l accessed) *Watershed Information Network*. U.S. Environmental Protection Agency. URL: http://www.epa.gov/win/.

World Resources Institute (2003) *Watersheds of the World – Primary Watershed Map, Water Resources Atlas*. Water Resources Institute, Washington, D.C.

1.3 Hydrologic Cycle

Black, P. (1996) *Watershed Hydrology*, Ann Arbor Press, Chelsea, Michigan.

National Aeronautics and Space Administration (NASA) (2004) *Global Hydrology Resource Center*. Global Hydrology and Climate Center, NASA. URL: http://ghrc.msfc.nasa.gov/.

1.6 – 1.8 Watershed Assessment, Planning, Management

Center for Watershed Protection (2006a accessed) *Watershed Assessment*. URL: http://www.cwp.org/tools_assessment.htm.

United Nations Educational, Scientific and Cultural Organization (UNESCO) (2006a accessed) *Flow Regimes from International Experimental and Network Data (FRIEND)*. International Hydrological Programme. URL: http://typo38.unesco.org/en/about-ihp/ihp-partners/assessment.html.

UNESCO (2006b accessed) *Water a shared responsibility*. The United Nations World Water Development Report 2. URL: http://www.unesco.org/water/wwap/wwdr2/table_contents.shtml.

UNESCO (2006c accessed) *World Water Assessment Programme*. URL: http://www.unesco.org/water/wwap/description/index.shtml.

U.S. EPA (2005) *Draft Handbook for Developing Watershed Plans to Restore and Protect Our Waters*. Report EPA 841-B-05-005, Office of Water, Washington, D.C.

WCED (World Commission on Environment and Development) (1987) *Our Common Future*, Oxford University Press, Oxford, UK.

1.9 Watershed Restoration

Center for Watershed Protection (2006b) *Watershed Restoration.* (Accessed 2006) URL: http://www.cwp.org/restoration.htm.

Maryland Department of Natural Resources (2006 accessed) *Watershed Restoration Action Strategies.* URL: http://www.dnr.state.md.us/watersheds/wras/index.html.

1.10 Organization Issues – Case Study Massachusetts Watershed Initiative

Massachusetts Department of Environmental Protection (MA DEP) (2002) *Massachusetts Watershed Initiative Program.* MA Department of Environmental Protection, Worcester, MA.

1.11 The U.S. Clean Water Act

Adler, R.W., Landman, J.C. and Cameron, D.M. (1993) *The Clean Water Act 20 Years Later,* Island Press, Natural Resources Defense Council, Washington, D.C.

U.S. EPA (2003) *Introduction to the Clean Water Act.* U.S. Environmental Protection Agency. URL: http://www.epa.gov/watertrain/cwa/.

1.12 Total Maximum Daily Load

Houck, O.A. (2002) *Clean Water Act TMDL Program: Law, Policy, and Implementation,* 2nd edn, Environmental Law Institute, Washington, D.C.

U.S. EPA (1997a) *Compendium of Tools for Watershed Assessment and TMDL Development.* Report EPA 841-B-97-006, U.S. Environmental Protection Agency, Washington, D.C.

U.S. EPA (1999a) *Protocol for Developing Nutrient TMDLs.* Report EPA 841-B-99-007, U.S. Environmental Protection Agency, Washington, D.C.

U.S. EPA (1999b) *Protocol for Developing Sediment TMDLs.* Report EPA 841-B-99-004, U.S. Environmental Protection Agency, Washington, D.C.

U.S. EPA (2001) *Protocol for Developing Pathogen TMDLs.* Report EPA 841-R-00-0002, U.S. Environmental Protection Agency, Washington, D.C.

U.S. EPA (2006c) *Examples of Approved TMDLs.* U.S. Environmental Protection Agency. (Accessed 2006) URL: http://www.epa.gov/owow/tmdl/ examples/

1.13 Pathway Analysis for Watershed Institutional Design

Agrawal, A. (2002) Common Resources and Institutional Sustainability. In *The Drama of the Commons,* (eds. E. Ostrom, T. Dietz, N. Dolsak, P. Stern, S. Stonich, and E. Weber), pp. 41-85, National Academy Press, Washington, D.C.

Black, P. (1996) *Watershed Hydrology,* Ann Arbor Press, Chelsea, Michigan.

Randhir, T.O., and Genge, C. (2005) Watershed-based, Institutional Approach to Developing Clean Water Resources. *Journal of American Water Resources Association.* **41**(2), 413-424.

Ostrom, E. (1990) *Governing the Commons: The Evolution of Institutions for Collective Action,* Cambridge University Press, Cambridge, UK.

U.S. EPA (1996c) *Watershed Progress: New York City Watershed Agreement.* Report EPA 840-F-96-005, U.S. Environmental Protection Agency, Office of Water, Washington, D.C.

U.S. EPA (2006j accessed) *Watershed Academy Web.* U.S. Environmental Protection Agency. URL: http://www.epa.gov/watertrain/watershedmgt/.

2

Land Use and Water Quality Issues

2.1 LAND USE IMPACTS ON WATERSHEDS

Land use activities vary among rural, suburban, and urban areas and can have significant impacts on natural processes and the functioning of a watershed system. Impacts on environmental quality can range from minimal influences to significant effects on water resources, public health, and biological diversity. Thus, the goal of sustainable watershed management is to balance natural resource use and protection to satisfy both short term and long term needs of the current generation, without impairing the quality of watershed systems available to future generations.

Watershed managers often face major challenges in managing water resource depletion and ecosystem changes resulting from land use activities. In the U.S., legislation such as the Clean Water Act of 1972 has been effective in achieving water quality improvements, but nonpoint source pollution remains a major challenge. Land use activities can generate pollution in the form of fertilizers, pesticides, sediment, solid waste, wastewater, pathogens, oil leaks, and other hazardous wastes. Runoff water from rainfall and snowmelt traverses various land uses and accumulates along the flow of the hydrologic cycle. The contaminated water discharges into downstream water bodies and sometimes can seep into groundwater systems. These pollutants have undesirable consequences on the economic, commercial, ecosystem, recreational, and aesthetic aspects of a landscape. In addition to water quality, the ecosystem and economic ramifications of pollution can be significant. This chapter provides an overview of the watershed implications of varying land uses and some approaches to mitigating these impacts.

2.1.1 Residential Activities

As the world's population increases, residential land use is expected to increase in watersheds throughout the world. Residential activities can have a significant impact on both surface and groundwater resources through the waste products generated by households. These include solid waste, household chemicals, lawn fertilizers, and building materials, which are released as nonpoint source (NPS) pollutants into the watershed ecosystem. Another major source of residential NPS pollution is human excreta from areas with poor sanitary facilities, septic system malfunctions, and leaking sewers.

2.1.1.1 Septic Systems

Septic systems are small-scale, biological treatment systems intended to handle residential wastewater (Figure 2.1). Household wastewater originates from residential toilets, sinks, washing machines, and dishwashers. The wastewater includes a wide variety of contaminants; among them are bacteria and viruses, grease, synthetic organic chemicals, suspended solids, solid waste, organisms that increase biological oxygen demand (BOD), and nutrients. Wastewater is collected in a tank, acted upon and broken down by microorganisms, then is flushed through a leach field that filters pollutants in the wastewater. These systems are intended to treat septic tank effluent through filtration and biochemical reduction and transformation of the pollutants (Winkler 1998).

When septic systems are properly designed and maintained through regular pumping of the accumulated waste products, they pose minimal problems to water quality. Septic system malfunctions may result from system overload, a low population of microorganisms, overflow of solids into the leach field, and clogging of the system. These overflows can pollute surface and groundwater resources of a watershed. The location of septic systems is also a critical factor influencing the level of contaminants in a watershed system, especially those systems located in sensitive areas associated with the lake fronts, riparian areas, coastal waters, and wetlands. Failing septic systems can pollute groundwater resources. Groundwater contamination of wells and aquifers can have serious impacts on public health and degrade freshwater supplies in a watershed. Contaminated groundwater is difficult and expensive to treat, and in some cases could be infeasible to restore.

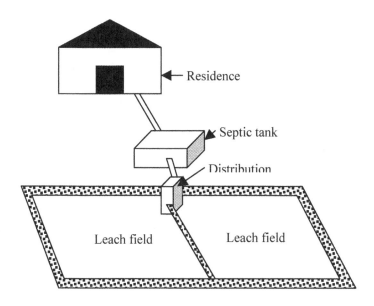

Figure 2.1 Layout of a septic system

Septic systems also have some advantages. Because of their small scale, they are easily installed in households and smaller locations with less space. The treated water is returned to the groundwater system through leach fields and is not exported out of the watershed system. This purified wastewater can replenish groundwater supplies.

Management of septic systems requires householders to maintain and upgrade their systems in a routine manner. Thus, educating householders with septic systems is of vital importance to protect drinking water supplies and to the overall health of the watershed.

Problems and Impacts

Nitrates Raw wastewater originating from septic systems contains organic nitrogen (N), urea, and dissolved ammonium-N. Through a series of chemical changes in the septic system, ammonium leaves the system as nitrate-N. Septic systems that are malfunctioning can cause nitrate contamination of water resources, especially lakes, ponds, aquifers, and deep wells. Excess nitrates in drinking water can also be harmful to human health.

Phosphates Balanced amounts of nutrients are essential to the health and normal functioning of watershed ecosystems. Phosphorus is a nutrient that is necessary for plant growth; however, when it is present in aquatic environments in excess amounts, phosphates can cause nutrient enrichment resulting in excessive plant growth and eutrophication (See Chapter 3, Inland Water bodies). Phosphorus is present in many household cleaners and wastewater and undergoes little treatment in the septic tank. This nutrient is discharged into the soil below the leach field. Phosphorus is not generally a problem in well-functioning septic systems. However, when phosphorus saturates soils, it can be steadily released to the environment over a longer period of time.

Pathogens Bacterial and viral pathogens in human waste, including E. coli and hepatitis A, are present in wastewater leaking from improperly functioning septic systems. These pathogens can cause serious human health problems when they contaminate water bodies that are intended for drinking water and recreation. Human illnesses caused by pathogens can necessitate costly treatments and the restoration of affected water bodies. The release of pathogens into water resources is of particular concern in "contributing areas" for drinking water supplies and aquifer recharge.

Solutions

Structural

Many alternative technologies exist to accommodate the limitations of the traditional septic tank and leach field. The problems associated with septic systems in vulnerable areas; for example, those with high water tables, near water bodies and in shallow soils over bedrock, may be overcome through the addition of enhanced treatment components to the conventional septic systems. Technologies have been developed to deal with a range of problematic soil conditions that include allowing for smaller soil absorption areas, reducing isolation distances, and providing for nitrogen credits. New systems have also been developed with alternate soil absorption systems as substitutes for septic tanks (Winkler 1998).

Nonstructural

- Regulatory approaches to siting and design can be used to mitigate septic system problems. Many U.S. states have enacted regulations pertaining to the treatment and disposal of sewage in areas lacking central sewage systems. For example, Massachusetts promulgated "Title 5" of the MA State Environmental Code in 1995 that governs the siting and design of septic systems to perform optimally at given locations.
- Regulations that emphasize analysis of soil characteristics can be used to target potentially problematic soil conditions or properties that may

negatively impact treatment efficiency. These conditions may include seasonal high water tables, flow restricting layers, stratification, and excessive stoniness (Winkler 1998).

- Education in septic maintenance and operation can include:
  - Avoid the use of chemicals such as anti-bacterials that kill natural microorganisms that break down waste. Do not put paints and strong household chemicals such as bleach and drain cleaners into the septic system.
  - Avoid introducing materials into septic systems that cannot be broken down by microbial action, such as cigarette filters, napkins, paper towels, and sanitary napkins.
  - Compost food wastes instead of using the garbage disposal.
  - Avoid the use of detergents with phosphates.
  - Follow maintenance and upgrade procedures for the proper functioning of the system.

2.1.1.2 Toxic Chemicals and Hazardous Materials

Description

Chemicals occurring naturally in the environment, such as nutrients, are essential to the health and functioning of ecosystem processes. Impairment of the normal balance of chemicals can affect the life, health, and reproductive capacity of organisms. Chemicals that cause deleterious effects on living organisms are categorized as toxic. Since the 1940s, more than 70,000 synthetic chemicals have been manufactured for use by residential homeowners. Many of the products used every day in and around residential and commercial facilities can be hazardous if released into the environment through improper disposal, storage or handling that result in leaks and spills into the watershed environment. Common toxic chemicals and hazardous products include most household cleansers, lawn and garden pesticides and herbicides, gasoline and petroleum lubricants, batteries, and home construction materials such as solvents, adhesives, paints, stains and varnishes, asphalt, and acids.

Problems and Impacts

Spills and leaks Chemicals can leak into the soil profile during storage or use. Indicators of such spills or leaks are pooling, discoloration, evidence of sheen, or detectable odors. Contamination of soil and water bodies by chemical and hazardous materials can degrade water supplies, pose public health problems, poison wildlife, and disrupt ecosystem functions.

Groundwater chemical plumes Chemical contaminants reach groundwater systems through discharge from surface water bodies or through infiltration through soils. Once in a groundwater system, these pollutants can spread quickly as contamination plumes. These plumes are high concentration zones of chemicals that move slowly, following the groundwater flow regime. Plumes can vary in the size of their coverage and speed of expansion, depending on the type, amount, solubility, and density of the contaminant, and the velocity of the groundwater. One of the most common and serious impacts associated with groundwater contamination exists where the contaminant binds to soil particles and thus can persist in the groundwater system for a long time. These contaminants are nearly impossible to remove completely or require prohibitive costs to thoroughly clean them up. Contaminant spills are often easier to detect than leaks, since they normally occur during use or application. In either case, both spills and leaks may or may not be evident visually. As a result of the groundwater plumes, some drinking water supplies become unsafe for human consumption, requiring closure or costly remediation.

Solutions

Structural

Collect and store chemicals and hazardous materials in labeled, watertight containers, in dry storage areas, and in accordance with manufacturer's recommendations.

Surfactant flushing (Figure 2.2) is a method that saturates a spill area with a soil detergent that settles the pollutant in an area outside of the aquifer.

Nonstructural

- Institute strict inventory and housekeeping procedures and practices for all chemicals used.
- Provide user training, making sure chemicals and hazardous materials are used in accordance with manufacturer's instructions and in compliance with any local, state, or federal regulations.
- Have emergency cleanup procedures and materials in place with proper training for personnel using the chemicals to control and clean up spills and leaks quickly and effectively.
- Dispose of unused chemicals in accordance with state and local laws through community-sponsored hazardous waste collection days or at licensed waste disposal facilities.
- Wherever possible, encourage use of nontoxic products that are biodegradable and ecologically safe.

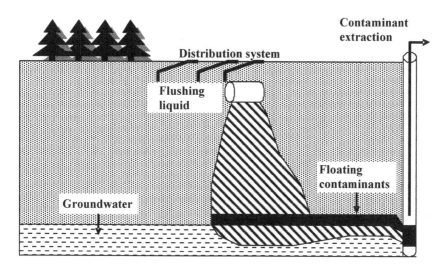

Figure 2.2 Surfactant flushing (Adapted from Ryan et al. 1995)

2.1.1.3 Pesticides

Insecticides, herbicides, fungicides, and rodenticides are broadly referred to as pesticides. Broad-spectrum pesticides are formulated to target a wide range of species. Selective, or narrow-spectrum pesticides target a single or small, well defined group of organisms. While pesticides respond quickly, reduce plant and animal diseases, increase food supplies, and reduce economic losses to farmers, they can have serious effects on water quality and human and ecosystem health. Homeowners use a variety of pesticides in their houses, gardens, lawns, and swimming pools. Pesticides are also found in a variety of household products that include paints, building materials, carpets, and mattresses. The U.S. Environmental Protection Agency reports that 92 percent of all households use pesticides, and that the average U.S. homeowner applies two to six times more pesticide per hectare than do farmers.

Problems and Impacts

Drinking water contamination Pesticides are a significant source of surface and groundwater pollution. If introduced into water supplies, these contaminants render the water unsafe for human consumption. Remediation may or may not be possible, and in either case it is expensive to treat contaminated waters.

Human health According to the U.S. EPA, more than 250,000 U.S. residents become ill each year from household use of pesticides through improper use or accidental poisonings. Pesticides are also identified as a major cause of accidental poisonings and death among children under the age of 5. The threats and human health impacts from pesticides predominate in developing countries because of their widespread use.

Fish and wildlife poisoning Pesticides are known to affect a wide variety of fish and wildlife. When released into aquatic environments either directly through spills, through runoff, or via wastewater, pesticides affect aquatic and terrestrial organisms by poisoning the food web and disrupting the ecosystem. Bioaccumulation of pesticides in organisms increases as the pesticides move up the food chain. Animals in higher trophic levels contain greater amounts of toxins in their tissue than those in lower trophic levels. Predators such as birds and mammals that are at the top of the food chain are most often subject to lethal levels of these pesticides, resulting in sickness, death, or reproductive failure.

2.1.1.4 Lawn and Garden Fertilizers

Description

Fertilizers applied to lawns and gardens can enter into runoff water and pollute downstream water resources. Nitrogen and phosphorus are two major components of most fertilizers and are major pollutants that contribute to the eutrophication of water bodies. When introduced into aquatic systems, nutrients can cause plant biomass to grow at a rate higher than that which aquatic ecosystems can naturally handle. The depletion of dissolved oxygen resulting from eutrophication can have serious effects on aquatic food chains.

Problems and Impacts

Eutrophication When phosphorus from fertilizers that are applied to lawns and gardens is washed into aquatic systems, especially lakes and ponds, the systems become overloaded and eutrophication occurs. Nitrogen is another nutrient important to plant growth that is commonly used in lawn fertilizer. Like phosphorus, when introduced into water bodies, nitrogen will promote excessive plant growth leading to eutrophication. Nitrogen loading is of particular concern in coastal watersheds, which are naturally low in nitrogen (nitrogen limited). Nitrogen is also converted to nitrates by bacteria and can leach into the water table and wells, where it causes contamination.

Lime effects on pH Some landscapes have acidic soils, which are not hospitable to grasses, plants, and shrubs requiring alkaline soils. Lime is often applied to lawns and garden soils to increase the pH (reduce acidity). Runoff from these locations can also change the pH in water bodies. Each aquatic species has specific tolerance ranges for pH. An increase in pH by the addition of lime can be intolerable to sensitive aquatic species.

Solutions

Structural

- Maintain or establish riparian buffers using native vegetation that is tolerant of soil acidity.
- Plant vegetation (filter strips) downstream of disturbed areas that can filter pollutants from runoff water before it reaches a water body.

Nonstructural

Educate the public and regulate fertilizer and lime applications with respect to the following:

- Regulate application of lime to soils and lawns along riparian areas.
- Protect naturally occurring leaf litter and vegetation that serve as a natural filter for fertilizers.
- Limit the size of lawns to reduce the need for lime and fertilizers.
- Avoid over-application of fertilizers, and use an optimal rate of fertilization with proper placement in the soil.
- Regulate timing to avoid application prior to precipitation events or during windy conditions.
- Nitrogen fertilization rates should be based on soil tests and recommendations for the particular lawn.

Case Study

Watershed Name: Lake Peipsi/Chudskoe-Pskovskoe Basin
Location: Border of Republic of Estonia and the Russian Federation
Major Problem: Pollution and eutrophication
Approaches: Water treatment plant in Tartu city. Integrated water resource management is being developed to coordinate the efforts of the two countries.
Information:
http://www.unesco.org/water/wwap/wwdr2/case_studies/pdf/lake_peipsi.pdf

---

2.1.1.5 Excessive Water Use

Description

As population increases at a rapid pace, the use of water by human beings is a major cause of the depletion of surface and groundwater supplies. In many regions of the world, excessive water use for residential purposes has become a serious issue - seasonally and, in some regions, year round. Measures to conserve and increase efficiency of water use can be achieved through public conservation education programs, government regulations, and incentive programs to adopt technologies. These measures can be a part of a watershed management plan.

Problems and Impacts

Water shortage Unmanaged water use can lead to rapid depletion and increase water scarcity. This can have serious consequences on the regional economy, ecosystems, and access to water.

Aquifer depletion Communities that rely on aquifers for residential water supplies can deplete the groundwater supplies if the resource is not managed well. The aquifer depletion can result in drawdown and land subsidence.

Surface water drawdown Surface waters such as lakes and rivers that communities rely upon for water supplies can be subject to seasonal variability in flow regime and availability. In association with water shortages, the flora and fauna of aquatic systems can suffer detrimental effects. The consequences of reduced water levels can include increased temperature, changes in dissolved oxygen (DO), and loss of habitat quality.

Solutions

Structural

Common recommendations to improve residential water use efficiency:
- *Locate and fix leaks* in the plumbing to reduce water loss.
- Use *low-flow plumbing* as a one-time investment to conserve water. This plumbing can be low cost and lead to substantial water savings in the long run.
- Install *pressure reduction valves* to reduce household water pressure in either municipal or private well water systems.

- Use *low-flush toilets* to minimize water use per flush. Low-flush toilets may use 1.6 gallons or less compared to 3.5 to 5 gallons or more in conventional toilets.
- Insert *toilet displacement devices* to reduce the amount of water used per flush.
- Install *low-flow showerheads*, low-cost apparatus that can save a substantial amount of water per family.
- Use *faucet aerators* to reduce the water use at a faucet without reducing flow. Entrained air forces water to flow in small droplets.
- *Reuse gray water.* Gray water is used domestic water from kitchen sinks, laundry and dishwashers. It can be disinfected and used for gardening, lawn maintenance, and other non-consumptive uses.
- Reduce *landscaping irrigation.* A homeowner can conserve water in lawn maintenance and landscaping by using plants that require less water, irrigating when evaporation loss is low, using low-precipitation sprinklers and drip systems, or xeriscape landscapes.

Nonstructural

- *Water pricing* can be used to provide incentives for water conservation. Metering of water use is a valuable way to monitor usage and develop appropriate pricing schemes.
- *Education* about water use practices can change water use habits to use water more efficiently and can result in decreased water use in the home. Examples include:
- Turning off the faucet when brushing teeth or shaving; using correct water levels in the washing machine to match the size of the load; watering gardens and lawns early in the morning or evening to reduce water loss from evaporation; turning off the hose between rinses when washing the car; and sweeping walkways instead of hosing them down.
- Water rationing, regulating the retrofitting of plumbing equipment, changing construction codes, and water reuse regulations can be used to conserve water.

Case Study
Watershed Name: Adour-Garonne Basin
Location: Southwest France, Europe
Major Problem: Low water levels
Approaches: Stakeholder involvement and allocation rules through planning tools such as strict low-water target flow (DOE) and low water management scheme (PGE).
Information: http://unesdoc.unesco.org/images/0014/001459/145925E.pdf

---

2.1.2 Municipal Sources

Municipalities are urbanized areas characterized by significant alterations in natural habitats. Municipal operations can have significant impacts on the environmental quality of watersheds. Besides altering habitats and species composition, municipal development can alter the natural topography and drainage patterns in a watershed. By diminishing the natural pollution-filtering capacity of vegetation and soils, urbanization indirectly contributes many forms of pollution. Stormwater management, wastewater treatment, impervious cover, combined sewer overflows, landfills, brownfields, and industrial pollution are some critical water quality concerns in municipalities.

2.1.2.1 Stormwater

Description

Stormwater runoff is a natural occurrence resulting from a precipitation event. The runoff is enhanced by increased development, which contributes higher flows of runoff and snowmelt to streams, rivers, lakes, and coastal waters. Human activities associated with urbanization affect stormwater runoff through the alteration of natural drainage patterns. Impervious surfaces impede infiltration of rainwater into the soil. Thus, water that contains contaminants bypasses the natural filtration processes that occur in the soil profile, and increased pollutants are added to water bodies. These contaminants degrade the overall environmental quality of watershed ecosystems. Ecosystem stress causes a decline in aquatic biodiversity and a reduction in species that cannot survive under altered conditions. In addition, human benefits from water resources are also diminished due to beach closures, public health concerns, and bans on the consumption of shellfish and other seafood (U.S. EPA 1996a).

Stormwater picks up and deposits pollutants and debris as it travels across developed areas. This runoff can contaminate waterways before there is a chance for filtration by natural vegetation and soils (U.S. EPA 1994). In the U.S. the Clean Water Act - Stormwater Phase One Rule (1990) requires municipalities with populations greater than 100,000 and construction sites greater than five acres to abide by discharge permit restrictions. The rule is designed to control stormwater runoff through the National Pollutant Discharge Elimination System Stormwater Permit Program. The Phase II rules (1997) apply to smaller cities and construction sites. In coastal municipal areas, the control of stormwater runoff containing sewage overflows is critical because of its impact on recreation and economically important aquaculture and coastal fisheries.

Problems and Impacts

Runoff contamination In municipal areas, stormwater runoff lifts and carries a variety of pollutants, such as solid waste, trash, and petroleum compounds from fueling stations, automobiles, and commercial vehicles This runoff can contribute human and industrial waste products to combined sewage overflows.

Erosion and sedimentation Erosion in municipal areas is the displacement of soils and substrates through urbanization and the creation of rivulets, gullies and erosion channels. Sedimentation results in turbid runoff waters and affects the quality of nearby streams. Sedimentation causes a number of detrimental effects to aquatic systems. Controlling stormwater runoff is essential in order to decrease sedimentation, reduce stream bank erosion and stream channeling, and reduce modifications to stream habitat in smaller streams. Runoff control also provides secondary benefits such as reducing the need for dredging and improving recreation (U.S. EPA 1998).

Solutions

The general goals of stormwater management are to maintain groundwater recharge and storage, reduce stormwater pollutant loads, protect stream channels, prevent increased overbank flooding, and to safely convey extreme floods. The Center for Watershed Protection (1998) recommends that watershed managers faced with selecting the stormwater BMPs for their watershed/subwatersheds should address the following issues and questions:
- What is the most effective mix of biological and engineering BMPs that can meet my subwatershed goals?
- What are the primary stormwater pollutants of concern (phosphorus, bacteria, sediment, metals, hydrocarbons, trash and debris)?
- Which hydrologic variables do we want to manage in the subwatershed (recharge, channel protection, flood reduction, etc.)?
- What are the best BMPs for removing pollutants?
- Which BMPs should be used or avoided in the subwatershed because of their environmental impacts?
- What is the most economical way to provide stormwater management?
- Which BMPs are the least burdensome to maintain within local budgets?

Structural

Retrofitting is the process of modifying existing surface water structures to control flooding. The objective is to improve water quality through hydrological control and pollutant removal. A number of retrofit techniques can be used, depending on their placement within the storm drainage network.

Selected examples of retrofitting (U.S. EPA 1994) are listed below:
- *Source retrofits* are techniques that attenuate runoff and/or pollutant generation before contaminants enter a storm drain system; e.g. reducing impervious areas and using pollution prevention practices.
- *Open channel retrofits* are installed within an open channel below a storm drain outfall, such as an extended detention shallow marsh pond system.
- A *natural channel retrofit* can be used depending on the size of the channel and the area of the floodplain.
- An *off-line retrofit* involves the use of a flow-splitter to divert the first flush of runoff to a lower open area for treatment. This technique is used when land is available for off-line treatment.
- An *existing BMP retrofit* focuses on improving the pollutant removal efficiency and/or capacity (ability to detain flow) of existing BMPs.
- An *in-line retrofit* can be used where there are space constraints that prevent the use of diversions to treatment areas.

Urban BMPs are generally grouped into the following categories: detention basins, retention/infiltration devices, vegetative controls, and pollution prevention. Other BMPs include sand filtration systems, underground sand filters with multiple chambers, filter strips, surface sand filter systems, and peat/sand filtration systems. Low Impact Development is a method to mitigate stormwater impacts of urban development through incorporation of better site design and environmentally sensitive features (U.S. EPA 2000).

Selected stormwater BMPs:
- *Bioretention* uses soils and plants to remove pollutants from stormwater runoff.
- *Catch basin cleaning* Catch basins retain some sediment from stormwater. In addition to sediment, materials such as decaying debris and solids are retained within the catch basin. Regular cleaning of catch basins can reduce loading of suspended solids, and oxygen-demanding substances into the surface waters.
- *Covering storage areas* by using roofs and sheds to store raw materials, products, equipment, process operations, and waste material that can contaminate stormwater.
- *Dust control* methods reduce the surface and air transport of dust caused by industrial and construction activities and prevent dust discharge into

stormwater. Regular street sweeping can be used to prevent solids and sediment from entering into runoff water.

- *Flow diversion structures* collect and divert runoff away from industrial areas and/or carry polluted waters to treatment facilities.
- *Infiltration drainfields* are designed to aid in stormwater runoff collection and subsequent infiltration into subsoil. *Infiltration trenches* can be used to remove suspended solids, particulate pollutants, coliform bacteria, and some soluble forms of metals and nutrients from stormwater runoff.
- *Porous pavement* is a specially designed pavement that is pervious to stormwater. This type of pavement reduces the speed and volume of runoff from a site and aids in the filtration of potential pollutants.
- *Urban forestry* can be used to increase canopy cover that can intercept precipitation and reduce runoff.
- *Vegetative filter strips* of sod, temporary and permanent seeding, and maintenance of existing vegetation, can be used to control erosion and dust from land use operations.
- *Vegetative swales* are broad, shallow channels with dense vegetation on the sides and within the main channel. They can be used to trap particulate pollutants, promote infiltration, and reduce stormwater velocity.

Nonstructural

Permitting A stormwater permit program can be adopted that requires permits for certain municipal and industrial stormwater discharges.

- A *stormwater utility tax* based on the percentage of impervious cover can be used to charge for stormwater discharges.
- *Zoning* regulates land use activities in specific locations of the watershed and is an effective way to control development and protect water quality. *Down-zoning* is regulation to require a lower density. *Conditional zoning* allows certain activities only under specified conditions in order to protect water quality. *Overlay zoning* places additional zoning requirements on an area that is already zoned for a specific activity or use. With the use of overlay zoning, high pollution activities can be controlled in sensitive areas of the watershed.
- *Cluster development* emphasizes compact placement of new homes to reduce the diffuse spread of impervious area. The use of compact development can help to preserve open space.
- *Open space preservation* protects undeveloped or minimally developed land and creates buffer zones near water bodies using greenways or riparian areas.

> **Case Study**
>
> Watershed Name: Connecticut River Watershed
> Location: New Hampshire, Vermont, Massachusetts, and Connecticut in northeast U.S.A.
> Problem: Water quality
> Approaches: Riparian buffers, elimination of combined sewer overflows, reduced effects of contamination, and improved water quality data.
> Information:
> http://www.mass.gov/envir/water/connecticut/connecticut.htm

2.1.2.2 Wastewater

Description

Most urban and municipal areas use centralized sewage systems to collect and treat wastewater. A network of pipes is used to collect the wastewater and transport it to a treatment center via pumping stations (Figure 2.3). At the treatment plant, wastewater is treated to remove some of the contaminants before discharging it into rivers or coastal waters. Sewage treatment plants use one to three levels of treatment prior to discharge. The least expensive and most common treatment is primary treatment, in which solids are removed by straining and settling. This treatment removes some organic solids in the wastewater. The secondary treatment removes a major portion of the organic solids through the use of settling tanks and microorganisms. In the tertiary treatment, other contaminants (such as nitrogen and phosphorus) are removed. Before discharge, wastewater effluent is often disinfected, usually via chlorination, to kill harmful pathogens.

Problems and Impacts

Inadequate treatment Many communities either do not have centralized treatment plants, or have systems with failing infrastructures, insufficient capacity, or operate only at the primary treatment level.

Combined sewage overflows Combined sewer systems which carry stormwater and sewage to treatment plants through the same pipes, are remnants of out-dated infrastructure. During periods of intense rainfall, when the demands exceed the treatment system's carrying capacity, combined sewers are designed to overflow directly to nearby waters without treatment (Figure 2.3). Combined sewage overflows (CSOs) create water quality and health problems by adding total suspended solids, BOD, total nitrogen, orthophosphate, metals, and toxic volatile organics to receiving water bodies (U.S. EPA 1996b). CSOs impair economic activities (e.g. fishing and boating) in developed areas. In coastal areas, CSOs can result in the shutting down of economically important aquaculture areas, beach closures, and have harmful impacts on aquatic ecosystems.

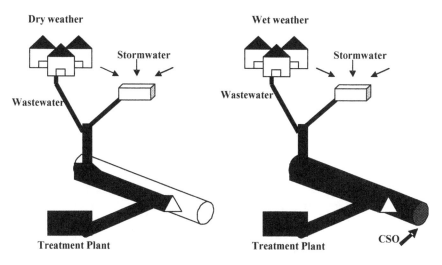

Figure 2.3 Wastewater treatment plant and combined sewage overflows

Sanitary sewage overflows A sanitary sewage overflow (SSO) results in raw sewage spills into basements or out of manholes and onto city streets, playgrounds and into streams. SSOs can occur when sewage pipes are blocked or restricted and result in breakage or leakage of the system. Pipe ruptures can also allow stormwater and groundwater into the system, resulting in increased volume of flow to the treatment facility. SSOs can be caused by inadequate sewer collection systems and poor maintenance of systems.

Solutions

Structural

- *Separation of storm and sanitary systems* is a solution to CSOs. This is often expensive and can be disruptive to implement.
- *Increase the treatment capacity* to treat the overflow. This is also expensive and does not solve the problem in the long run.

- *Reduce stormwater runoff* by increasing pervious cover in the watershed. This will reduce the incidence of excess overflows.
- *Community monitoring* can be used to quickly identify sanitary sewer overflows. Constant monitoring of the sanitary system for leaks and breaks can be part of the management objective.

Nonstructural

Incentive programs to reduce stormwater flows could be developed at a watershed scale to reduce overflows.

Education on treatment infrastructure and management is useful to increase public support.

2.1.2.3 Imperviousness

Description

Figure 2.4 Impervious cover increases in suburbanizing areas

An impervious surface is defined as "a hard surface area that either prevents or retards the entry of water into the soil mantle as under natural conditions prior to development and/or a hard surface that causes water to run off the surface in greater quantities or at an increased rate of flow from the rate of flow present under natural conditions prior to development" (U.S. EPA 1994). Impervious surfaces are characteristic of urban areas and impede the infiltration process (Figures 2.4 and 2.5). Vegetation and soils play a critical role in the hydrologic cycle by absorbing and storing water and filtering pollutants from runoff water. When vegetation is replaced and soils are covered with impervious materials, the ability of the soils to filter and purify contaminants contained in rain and runoff is diminished. The amount of impervious land in a watershed can be calculated by the following equation:

Impervious cover in watershed (%) = (Total impervious surface area / Total watershed surface area) * 100

Levels of impervious cover between 10 and 20 percent can create distinct water quality problems (U.S. EPA 1996a). The least developed urban areas, suburban residential districts, have an imperviousness rate around 20 percent. Highly developed business and commercial districts can have imperviousness rates greater than 50 percent. The primary effect of an increase in imperviousness is a sharp rise in runoff volume. The runoff travelling over impervious land acquires pollutants and discharges them into receiving waters. This leads to degraded water resources and has negative effects on downstream ecosystems.

All runoff from impervious areas need not contribute to stormwater flows. This is because some rooftops drain into lawns that allow infiltration. Thus, the Effective Impervious Area (EIA) is the percent impervious area in the watershed that is directly connected to the stormwater system.

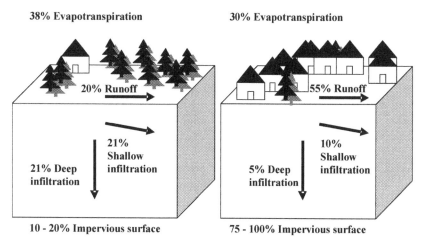

Data from Schueler and Holland, 2000

Figure 2.5 Hydrologic impacts of impervious cover

Problems and Impacts

Surface water runoff Flooding and the concentration of a large volume of stormwater runoff within a short time are common in areas with high imperviousness. Impacts of stormwater runoff include increased erosion where runoff comes in contact with pervious soils, increased sedimentation and bank erosion, and the deposition of pollutants into streams and lakes.

Roadway runoff - salt and sanding The use of salt and sand is common in areas which experience excessive ice and snow during winter time. Sodium chloride (salt) is a contaminant that is toxic to aquatic organisms, alters plant transpiration processes, and can prevent seasonal vertical mixing of stratified water. Sand adds to sedimentation, causing turbidity and increasing deposition in stream bottoms. Salt piles, which are exposed to wind and rain, can lose up to 10 percent of the volume of the uncovered pile, increasing salt and sediment content.

Fragmentation Impervious cover can fragment open space and available habitat in watershed landscapes. This fragmentation can impact sensitive organisms that need large patches that are not isolated.

Solutions

Structural

- Use pervious materials, such as trap rock, for driveways and roadways where possible.
- Limit or avoid large parking lots in areas where stormwater retention occurs naturally and in areas of slow flow and wetlands.
- Use porous pavement in parking lots and other urban infrastructures.
- Use stormwater retention systems that can temporarily store excess runoff.
- Cover and reduce leakage from storage facilities used for road salt. Use enclosed (roofed) structures constructed on flat sites, and cover salt piles (MA DEP 2006).
- Establish monitoring wells around salt/chemical storage facilities.
- Design and install drainage diversions for surface water runoff from storage structures.
- Incorporate pervious cover into development design to reduce runoff.
- Develop greenways or other types of corridors that connect patches that are isolated by urban development.

Nonstructural

- Develop regulation to avoid locating salt piles within public water supply watersheds.
- Create zoning policies to promote sustainable land use to reduce the impacts of impervious cover.

Case Study

Watershed Name: Rhine-Meuse Basin
Location: Transboundary basin in Europe encompassing Austria, Belgium, France, Germany, Italy, Liechtenstein, Luxembourg, the Netherlands, and Switzerland
Major Problem: Water quality and flood control
Approaches: Transboundary cooperation forum among countries in the basin.
Information:
http://unesdoc.unesco.org/images/0014/001459/145925E.pdf

2.1.2.4 Contamination from Metals and Hydrocarbons

Description

Metals are elements that can occur naturally in aquatic environments in low concentrations and have no adverse effects on human or animal health. When metal concentrations are elevated relative to background levels, adverse ecosystem impacts can occur. Once in the food web, these chemicals can bio-accumulate in organisms. In addition to being dependent upon the metal's speciation, the toxic effects of metals depend upon synergistic effects, the sensitivity of the organism, and rates of bio-magnification.

Problems and Impacts

Mercury cycles in the atmosphere, soils, and waterways, through a series of complex chemical and physical transformations. In air, water, soils, plants, and animals most mercury exists in the form of inorganic salts and organic mercury (U.S. EPA 1998). Mercury as methyl mercury accumulates efficiently in aquatic environments and bio-accumulates as it passes through the food chain. Consequently, predatory fish can develop high concentrations in their flesh, reaching levels toxic for human consumption. Although not a widespread health concern, people who consume large quantities of fish that are known to accumulate mercury are at risk for mercury poisoning. Since developing fetuses are the most sensitive to the effects from methyl mercury, childbearing women are of greatest concern for mercury poisoning.

PCBs and PAHs Polychlorinated biphenyls (PCBs) and polynuclear aromatic hydrocarbons (PAHs) are synthetic organic compounds that are persistent in the environment and have serious potential to impair human health and wildlife. The level of pesticide toxicity and its effects depend upon a number of variables including the dose, exposure time, solubility, and life stage of exposed individuals.

Arsenic is a naturally occurring element that can enter into water through runoff and leaching. It persists in the environment and can cause severe health effects that range from allergic reactions to death. It is also known to cause cancer in the lungs, bladder, liver, kidney, and prostate gland.

Solutions
Structural
- *Treatment* of arsenic contaminated water.
- *Change the source* of water in arsenic prone areas

Nonstructural
- *Regulate arsenic treated wood* and use protective gear to reduce exposure to saw dust.
- *Provide incentives* to use cleaner sources of water through education and subsidies.
- *Regulate pumping* and over use of water in areas prone to contamination.

2.1.2.5 Petroleum Storage and Spills
Description
Spills from petroleum and petroleum products are a common source of water and soil contamination. Petroleum spills can cause widespread and long-term impacts in watersheds. Petroleum contamination occurs from three primary sources: faulty storage tanks or leaking pipelines, accidents involving tank trucks and railroad cars, and spills or leaks at fueling stations and on roadways (Baehr and Corapcioglu 1984). Leaking underground tanks cause serious groundwater and drinking water supply contamination. Legislation to inspect storage tanks and mandatory replacement of leaking systems are often critical in preventing tank leakage. Petroleum spills are common at construction sites, industries, and fueling stations. Petroleum products such as oils and lubricants

can also leak from automobiles and other machinery. Direct introduction of oil products into stormwater sewers is through improper disposal.

Problems and Impacts

Leaks from underground tanks Water resources can be seriously impaired by petroleum contamination, as small amounts of gas and oil can render large supplies of water unusable. Petroleum spills at fueling stations and petroleum lubricants from automobiles are significant sources of nonpoint source pollution. Even though gasoline leaks from underground tanks at fueling stations are small compared to those from refineries and interstate pipelines, their frequency and widespread distribution are of concern to policy makers (Baehr and Corapcioglu 1984). When significant amounts of petroleum products contaminate soils, they spread rapidly downward through the unsaturated zone (Figure 2.6). Once at the capillary zone, the pollutant spreads into the water table. The nature of its distribution is determined by the total quantity of oil infiltrated, the infiltration rate, the hydraulic gradient, and the permeability of the soil (Baehr and Corapcioglu 1984). Leaks are often visible on impervious surfaces as sheens and discolored spots.

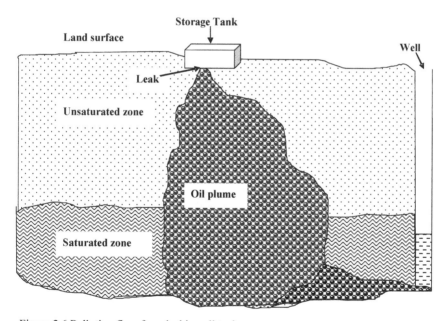

Figure 2.6 Pollution flow from leaking oil tank

Solutions

Structural

- The design, placement, maintenance, inspection, and replacement of underground storage tanks must be tested and applied to the local characteristics of the watershed.
- Installation of leak detection equipment for piping and storage systems can be useful in responding quickly to any leaks.
- New installations can use secondary containment tanks with an interstitial space leak detection system.
- Above ground non-corrosive bulk storage tanks can be used for non-explosive liquids.

Nonstructural

- The design, placement, maintenance, inspection, and replacement of underground storage tanks must comply with local, federal, and state laws.
- Develop regulations on spill prevention for each facility.
- Create incentives to use recycling facilities for used motor oil.

2.1.2.6 Brownfields

Description

Abandoned industrial facilities are potentially hazardous sites that are point sources of contaminants in a watershed. Old factories and processing stations often contain hazardous materials that are a continuous source of pollutants and pose health hazards to communities near them. Brownfields are defined by the U.S. EPA as "abandoned, idled, or under-used industrial and commercial facilities where expansion or redevelopment is complicated by real or perceived environmental contamination" (U.S. EPA 2001). Redevelopment and restoration of brownfields are often avoided due to the high costs of cleanup compared to development in less expensive areas on unpolluted land.

Problems and Impacts

Groundwater contamination Brownfields often contribute to groundwater pollution through old and leaking underground storage tanks. Most of the older facilities frequently have no records of storage tanks and disposal sites, making the location and removal process very difficult.

Public health concerns The presence of contaminants in brownfields can pose a serious threat to public health. The extent and the level of this threat depend on the location, type, and degree of contamination present at the site.

Solutions

Structural

Environmental cleanup of the area often involves detailed assessment and remediation methods. Cleanup often involves removal of the source of contamination, containment or treatment of the contaminated material, and monitoring of changes in the system.

Nonstructural

Flexible rules to allow parties to clean up efficiently, offering liability relief for contamination before land is purchased, and financial incentives are some policies used to facilitate cleanup of brownfields. Massachusetts in the U.S. has developed such rules and guidelines to help facilitate the cleanup of brownfields.

Federal grant programs, such as the "Superfund" in the U.S., are a part of the Comprehensive Environmental Response, Compensation, and Liability Act (CERCLA). The programs are designed to prioritize Brownfield sites and provide federal assistance for remediation and cleanup.

2.1.2.7 Landfills

Description

Private and municipal landfills are disposal sites for residential, commercial, and industrial wastes that can cause contamination problems in a watershed. The primary concerns result from the toxic and hazardous nature of wastes disposed in landfills and their leachates. Sources of leachate include residential and municipal solid waste, the degradation products of solid waste, illegally disposed hazardous wastes, and hazardous wastes from small quantity generators that continue to be legally disposed in landfills (Brown and Donnelly 1988). In addition to water quality problems, landfills are a source of health concerns due to nuisance animals, disease transmission, and diminished air quality. The concentration of effluent, discharge rates and the bed radius determine infiltration properties and the potential threats from a landfill (Ostendorf et al. 1984).

Problems and Impacts

Leachate and groundwater contamination Contamination from leachate occurs as precipitation percolates through decomposing wastes in landfills (Figure 2.7). The leachate moves in subsurface flows following the direction of groundwater flow (MA DEP 1993). Because the contamination from a landfill occurs below ground, the problem cannot readily be readily detected. Toxic chemicals and hazardous wastes that can contribute to landfill leachate include toxic cleansers, pesticides and fertilizers, paints, solvents, waste oil, and metabolites from plastics, paints, and pharmaceuticals. The impacts are serious, resulting in health problems or rendering wells unusable.

Surface water runoff Surface water contamination from landfills is more easily detected through visual observation of leachate runoff and from biological symptoms such as plant die-offs and wildlife poisonings, or through chemical testing and monitoring.

Nuisance animals Wildlife and domestic animals are attracted to landfills to scavenge for food. Sea gulls are often a significant problem in many landfills. Increasingly, coastal gulls are moving inland to feed at landfills, resulting in the contamination of water bodies they choose for resting or nesting, through fecal matter, and other attached contaminants. Dogs, skunks, foxes, bears, and other wildlife are common visitors at dumps and landfills.

Public health issues, disease transmission Food and solid wastes disposed at landfills often contain pathogens, which can be transmitted to humans through leachate, surface runoff, and nuisance animals.

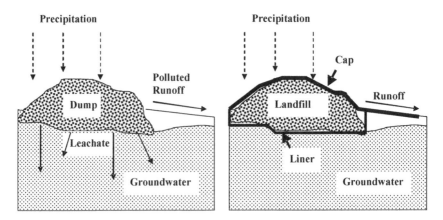

Figure 2.7 Landfill designs

Solutions

Structural

- *A minimum groundwater protection system design* involves combining a subgrade layer, a composite liner, a drainage protection layer, a leachate collection system and a pumping system (Figure 2.7).
- *Liner systems* prevent leachate from reaching groundwater by collecting leachate for treatment and disposal. Liners should be designed as closed systems and provide an effective hydraulic barrier during the active life, closure, and after closure of the landfill.
- *A subgrade design* provides a proper foundation for the landfill.
- *Low permeability soil/admixture layer* acts to minimize the movement of leachate into the substrate and groundwater.
- *Flexible membrane liners* provide an additional barrier layer in the liner design to prevent contamination from leachate.
- *Liner sideslope design* provides stability to landfill sidewalls upon compaction.
- *Drainage/protection layer* serves as a high permeability pathway through which leachate can travel to collection pipes.
- *Final cover systems* minimize groundwater pollution and costs associated with collection and disposal of leachate (MA DEP 1993).

Nonstructural

- Quality Control/Quality Assurance Programs can be used to regulate landfill construction, design specifications, installation of a liner or final cover, inspection, sampling and analysis, and documentation.
- Environmental monitoring programs can be used to monitor the quality of groundwater and surface water, leachate, landfill gas production, hazards, and precipitation effects using perimeter probes and monitoring wells.
- Landfill assessments can determine the impact of the landfill on groundwater, surface water, and air quality by qualitatively and quantitatively characterizing the nature and extent of the contamination and assessing the associated risks to public health and the environment.

2.1.2.8 Industrial Pollution

Description

Industrial pollution is often point source pollution caused by industries that contaminate air and water resources. Some industries with diffuse sources of pollutants can produce nonpoint source pollution. Significant detriments to water quality exist through wastewater discharges, storage leaks, and spills involving chemicals and hazardous substances.

Problems and Impacts

Wastewater Toxic industrial waste products include PCBs, arsenic, lead, mercury, and chlorine. Conventional industrial pollutants include oil and grease, total suspended solids, and BOD (EPA 1996b), as well as commonly used chemicals such as alkaloids, ammonia, chromium, cyanides, fluorides, hydrocarbons, mineral acids, sulfides, zinc, and phenols. Businesses in strip malls, industrial parks and areas that are not served by municipal sewage systems often resort to disposing their wastes in shallow wells or in septic systems designed to treat sanitary wastes (EPA 1996b).

(More discussion on wastewater can be found in Section 2.1.2.2.)

Chemical and hazardous waste storage spills and leaks from industries containing chemicals and hazardous wastes can occur during use on site, and during transportation.

Air pollution Atmospheric release and transport of industrial pollutants form a significant amount of pollutants returning to watersheds as downwind fallout. More than 23 million tons of nitrogen are released into the atmosphere each year, and about half of the nitrogen compounds are emitted from fossil fuel burning plants, vehicles, and other sources. The nitrogen is deposited onto watersheds, much of it falling directly onto water bodies (EPA 1998). The atmospheric source of nitrogen can be significant in some watersheds. Acid rain, caused by the emission of sulfur oxides from coal-burning power plants, contributes to increased acidity of aquatic environments rendering many inland water bodies and wetland areas unsuitable for aquatic life.

Solutions

Structural

- *Onsite treatment* is one option industries can consider to reduce pollutant loading into the watershed.
- *Constructed wetlands* and *vegetative filter strips* can be used to avoid contamination downstream.

Nonstructural

- *National discharge standards* for toxic and conventional pollutants can be used to regulate pollution.

- *Incentives* to adopt onsite treatment technologies can be used to encourage pollution reduction.
- *Market-based approaches* such as trading in nutrients and air emissions can be cost effective.

Case Study

Watershed Name: Artols-Picardy Basin
Location: Northern France, Europe
Major Problem: Water pollution and groundwater depletion
Approaches: Groundwater abstraction and pollution charges. Considerable reduction in pollution and groundwater abstraction achieved.
Information:
http://unesdoc.unesco.org/images/0014/001459/145925E.pdf

2.1.2.9 Excessive Groundwater Use (Regional)

Description

Communities reliant upon aquifers as primary or secondary sources of water supply must assess potential impacts of excessive use or drawdown on water resources. The yield of wells varies considerably, depending on factors of location such as the geologic and hydraulic characteristics of the aquifer (e.g. consolidated vs. unconsolidated materials). Other factors include the location of the pump, and increases in groundwater development from the upland watersheds to the lower basins and floodplains. Under natural conditions, the hydrologic cycle can maintain water balance equilibrium, and good management can assure a safe yield from the aquifer. The safe yield is often defined as the amount of groundwater that can be withdrawn annually without producing excessive drawdown; i.e., without impairing the water balance. Overdrawing groundwater may result in increased pumping costs and environmental problems such as saltwater intrusion and land subsidence. Ideally, groundwater should be managed over long periods of time so as to assure no change in storage. (Brooks et al. 2003).

Problems and Impacts

Sinkholes Rapid expansion of groundwater withdrawal can reduce water pressure to the point where limestone or other soil layers collapse to form sinkholes. This can result in property damage, loss in land value, and potential injury to land users.

Saltwater intrusion In coastal watersheds, reduced water pressure in freshwater aquifers creates a hydraulic gradient that favours movement of saltwater into the aquifer. This condition renders freshwater unusable and results in a loss of supplies. Saltwater intrusion can also result in corroded pipes and degraded infrastructure of coastal watersheds.

Subsidence and compaction When groundwater pressure is reduced in an aquifer that lies beneath thick, compressible clays and silts, the aquifer material is compressed under the weight of the overlying sediments. This compression can cause sinkholes, destroy drinking water wells, change the hydrology of springs and streams, and damage buildings and other infrastructure.

Depletion Wells may run dry due to excessive use, usually during prolonged dry periods or droughts. Dry wells can cause serious economic losses and potential loss of life. When the water table is deep enough, it is possible to alleviate the condition by increasing the depths of wells. However, this is very expensive and may lead to further depletion of the water table.

Household excess water use Inefficient household water use is a large source of wasted freshwater. Using efficient plumbing such as low flow showerheads and smaller capacity toilets can increase water conservation. For example, washing cars over an open lawn instead of a driveway allows the water to infiltrate the ground instead of entering the sewer system. (See Section 2.1.1, Residential Activities, for more information.)

Solutions

Structural

Implementation of water conservation methods is an indirect way to reduce overuse and exploitation.

Costly treatments or substitution of alternate water sources are needed to maintain water supplies impacted by salt-water intrusion. In some situations, costly artificial recharge methods, which add freshwater to the aquifer through wells, can be employed to correct salt-water intrusion.

Nonstructural

Regulation of the rate of pumping and level of interference (spacing of wells) are important to achieve sustainable use of groundwater resources.

A reduction in energy subsidies can be used to minimize over pumping in areas where energy costs are highly subsidized.

Metering and pricing methods can be used to reduce excess pumping in sensitive aquifers.

Case Study

Watershed Name: Romwe Catchment
Location: Southeastern Zimbabwe, Africa
Major Problem: Land use impacts on groundwater
Approaches: Multidisciplinary participatory research and decentralized decision making.
Information:
http://www.fao.org/ag/aGL/watershed/watershed/papers/papercas/paperen/case19en.pdf

---

2.1.2.10 Right of Way (R.O.W.) Maintenance

Description

Municipalities and towns maintain right-of-ways (ROWs) along roadsides in order to maintain town-owned infrastructure such as water and power lines. The same is true for utilities and commercial enterprises, which provide cable, electricity, natural gas, and rail line services to consumers. Developing and maintaining these ROWs can have detrimental impacts on ecosystems and watersheds.

Problems and Impacts

Electric power transmission Environmental impacts can be caused by the construction, operation, and maintenance of transmission lines. Electric power lines must be installed and maintained along roadways, which require maintenance and trimming of roadside vegetation. Transmission lines are primarily overland systems constructed to span or cross wetlands, forests, streams and rivers, and near shore areas of lakes, bays, etc. The construction of ROWs through forest areas can result in forest fragmentation that affects wildlife populations and ecosystem species composition (see Chapter 5, Biodiversity and Ecosystem Health). The use of herbicides to control vegetation under transmission lines can pollute aquatic systems and negatively affect wildlife and human health. Some herbicides are linked to human cancer, birth defects, and changes in hormonal systems.

Railways The construction and maintenance of railways result in many of the same environmental impacts as transmission lines, most importantly fragmentation, and herbicide use.

Roadways Construction of roadside ROWs can cause erosion and contribute sediment to area water bodies. Construction related to road repair can increase pollution and onsite erosion.

2.1.3 Construction

As natural landscapes are converted to human use, the changes can have significant effects on watershed ecosystems. Construction activities have numerous effects on both terrestrial and aquatic environments. Some effects of construction include: tree and vegetation removal which destabilizes soils and results in on-site erosion and off-site sedimentation; excavation, grading, and filling operations which lead to alteration of topography and drainage patterns; chemical spills and hazardous waste pollution which contaminate soil and water sources; sedimentation of waterways, and the alteration of stream morphology that disturb aquatic ecosystems. Many of these effects result from the initial land clearing process during construction.

2.1.3.1 Erosion and Sedimentation

Description

Soil erosion is initiated by precipitation that lifts soil particles, moves them down slope along with runoff and deposits them into water bodies (Figure 2.8). Runoff that gains sufficient velocity cuts small rills and gullies on the land surface, adding a sediment load to the runoff water. As the velocity, volume of water and soil load that is being carried increases erosive effects can be harder to control.

Sediment loads in water and wind are deposited on land or in water, depending on the speed of the transporting medium, the weight of the soil particles, and any physical impediments to flow. Smaller soil particles are capable of staying suspended in water and wind for long periods of time and over great distances.

Problems and Impacts

Soil erosion While all soils are susceptible to water and wind erosion, specific factors enhance the erosion process. Major factors that influence erosion potential are rainfall erosivity, soil type, land slope and length of flow, land use type, and conservation practices adopted. Soil erosion can have serious impacts on receiving water bodies by altering water quality and harming aquatic life.

Figure 2.8 Soil erosion undercuts a slope, removing vegetation and exposing roots

Stream effects, sedimentation Large storms can result in a large volume and high velocity of runoff that are capable of moving great quantities of soil particles. Similarly, powerful and persistent winds can dislocate significant loads of small soil particles as dust. Sediment washed into receiving waterways can increase turbidity.

Solutions

Structural

The most important measure that can be taken to control erosion and sedimentation on construction sites is to employ best management practices that are suited for the location. One example is to use silt fences around construction sites to avoid soil loss from the site.

2.1.3.2 Chemical, Petroleum and Hazardous Waste Spills

Description

Construction activities often require dealing with large equipment and hazardous construction materials. When chemicals, petroleum, and hazardous materials are improperly handled or stored, spills and leaks can contaminate soil and water resources.

Problems and Impacts

Spills and leaks Equipment fueling spills, leaking chemical storage drums, and litter from hazardous materials can pollute soils and be carried by erosive forces into nearby waterways. Petroleum and chemical spills and leaks that are not contained and properly cleaned up can leach into groundwater supplies.

Solutions

Nonstructural

Communities adopt erosion and sediment management practices for construction sites to reduce potential impacts from construction activities.

2.1.4 Mining Operations

The mining and processing of minerals and metals for industrial, residential and commercial uses provide society with important functions. The mining industry removes large amounts of rock and soil from the earth each year, and regulation of these activities is limited. The mining process can create detrimental environmental impacts that include land disturbance, erosion, water pollution, and air pollution. These impacts can affect both human and ecosystem health.

Mining can necessitate the excavation of the earth at the surface level (e.g. strip mining), subsurface (e.g. open pit mining), or be conducted through deep extraction, as occurs with gas and oil. Each of these techniques has unique environmental impacts, and the specific impacts can vary according to the material that is being mined.

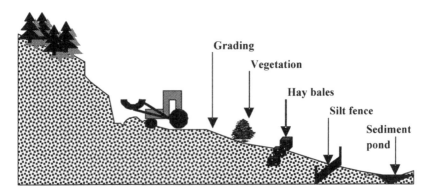

Figure 2.9 Surface mining BMPs

Problems and Impacts

Erosion and sedimentation Sand and gravel pit operations are capable of contributing large amounts of sediment to adjacent water bodies due to water erosion and wind displacement of smaller soil particles (Figure 2.9).

Groundwater contamination Sand and gravel are porous enough to transport soluble pollutants to underlying groundwater. Excavation near the recharge areas of public and private wells can have serious impacts on drinking water.

Surface scarring from surface mining can result in ecological disruption, especially if reclamation is not implemented. The piles of earth material in spoil heaps and tailings contribute to soil, water, and air pollution.

Effects on biogeochemical cycles Mining of mineral deposits that contain nitrate and ammonium ions for use as fertilizers can introduce large amounts of these nutrients into the nutrient cycles. Phosphate rock is often mined for use as commercial inorganic fertilizer and for detergents. Both of these mining operations can interfere with the natural functioning of both nitrogen and phosphorus cycles.

Solutions

Structural

The following BMPs can be used to mitigate the impacts of mining operations (Massachusetts Bays Program 1998):
- Monitor water flows entering and leaving mine sites to study the status and nature of contamination.
- Redirect uncontaminated water away from contaminated areas to avoid further contamination.
- Use catch basins, contour terraces, and cover crops to capture sediment, prevent erosion, and enhance aesthetic values in areas impacted by mining operations.
- Treat contaminated water using wetlands and other small treatment methods.
- Stabilize mining waste areas to prevent release of materials to streams.
- Maintain riparian buffer strips along streams.

Nonstructural

- Require a full environmental impact assessment of all mining activities.
- Adopt environmental standards for pollution and environmental degradation resulting from mining activities. Standards can be used to

regulate mining operations (e.g. excavation size and pollution mitigation measures) and restoration (e.g. slope restoration and establishment of vegetative buffers).
- Create mining bylaws to minimize impacts on the hydrologic cycle.

2.1.5 Agriculture

Figure 2.10 Animal Feeding Operation (AFO)

Agricultural activities such as livestock operations and crop production can have a significant effect on water quality and habitat in watershed ecosystems. These activities are nonpoint sources of pollutants affecting water quality. Livestock operations and crop production can increase nutrient loading, promote eutrophication in water bodies, introduce pathogens, and affect instream water flows. Excessive livestock grazing can alter natural habitats and species composition. Grazing can cause erosion and increased sedimentation, alter stream morphology, and cause severe damage to riparian areas. Soil tillage for crop production can make soils vulnerable to erosion and increased overland flow. Pesticides and fertilizers affect water quality and downstream habitat. In recent years, significant progress has been made through regulatory programs and implementation of best management practices to minimize watershed degradation resulting from agricultural activities.

2.1.5.1 Animal Wastes

Description

The U.S. EPA estimates that there are approximately 450,000 Animal Feeding Operations (AFOs) in the United States ranging from small livestock facilities to large production facilities (Figure 2.10). These generate animal wastes equivalent in magnitude to that produced by a medium-sized city (U.S. EPA 1998). Agricultural operations involving dairy cows, beef cattle, hogs, and chickens generate a number of pollutants which can significantly impact both surface and groundwater resources. In addition to the fecal and urinary wastes originating from livestock and poultry, animal wastes include process water (e.g. from milking parlors), and the feed, bedding, litter, and soil with which they become intermixed (U.S. EPA 1993). AFOs can be confined and are often called CAFOs (Confined AFOs). Storage from CAFOs and decomposing animal carcasses can contain oxygen-demanding substances such as nitrogen, phosphorus, organic solids, salts, and pathogens (bacteria, viruses and other microorganisms).

Two common problems associated with animal waste are the limited use of effective handling practices and the limited cost-effectiveness of regulations intended to restrict adverse effects of animal waste. Large volumes of liquid waste, lack of adequate manure storage facilities, lack of alternative uses of wastes, and failure to use management plans for handling wastes appropriately are specific problems facing AFOs (USDA 1998a). In addition to the problems connected with animal wastes, livestock open grazing operations can result in habitat degradation, loss of vegetation and productive soils, disruptions in species composition, and disturbance to water bodies.

Problems and Impacts

Nutrient enrichment Nutrients in animal wastes such as nitrogen and phosphorus are commonly disposed of and used as manure in crop production. Runoff from manure application to crops and leaching from storage facilities can contaminate both surface and groundwater resources. Leachate from livestock wastes stored in unlined manure pits can be detected in groundwater supplies. The best visual evidence of nutrient enrichment can be seen in farm ponds and slow-moving rivers as eutrophication caused by increased aquatic plant growth (see Chapter 3, Inland Water Bodies). Aquatic ecosystems can be impacted by contamination. Air quality can also be diminished by odors emanating from animal waste.

Pathogen introduction Animal wastes contain a variety of bacteria and viruses, many of which are pathogens. Diseases, which can be passed from animal waste to humans, include cholera, tuberculosis, typhoid fever, salmonella, and polio.

In addition, animal wastes contain harmful parasites, including Giardia and Cryptosporidium (USDA 1998a).

Solutions

Structural

A range of BMPs can be used to allow for the presence of animals in sensitive areas without adversely affecting water quality (USDA 1998a):

- Minimize runoff from standing, feeding, or watering sites.
- Make and follow a plan for waste management, including nutrient budgeting.
- Upgrade existing manure holding facilities.
- Build properly constructed new manure holding facilities.
- Apply wastes to land in accordance with BMPs for timing, application rates, etc.
- Properly store, dispose, handle and transport wastes.
- Test manure for plant nutrients.
- Site and/or locate pasture relative to access to streams and water.
- Calibrate manure spreaders.

Nonstructural

- Develop nutrient management regulations.
- Give incentives to farmers to implement nutrient management plans and adopt BMPs.
- Educate farmers on the impacts of nutrient loads on water bodies as a policy for voluntary compliance.

Case Study

Watershed Name: Loire-Brittany Basin
Location: Central France, Europe
Major Problem: Agricultural nonpoint source pollution
Approaches: Agro-environmental measures and nitrogen absorption program reduced impacts of agriculture on water quality. Financial incentives based on voluntary participation.
Information:
 http://unesdoc.unesco.org/images/0014/001459/145925E.pdf

2.1.5.2 Livestock Effects on Riparian Areas
Description

Figure 2.11 Farm animals with access to streams and ponds can severely impact water quality

The functioning of riparian areas is a result of complex interactions between soil, water, and biota. Proper functioning of the riparian ecosystem requires adequate stream-side vegetation to:

- Dissipate stream energy associated with high water flows (resulting in reduced erosion and improved water quality);
- Filter sediment and aid floodplain development;
- Support denitrification of nitrate-contaminated groundwater as it is discharged into streams;
- Improve floodwater retention and groundwater recharge;
- Develop root masses that stabilize banks against cutting action;
- Develop diverse ponding and channel characteristics to provide the habitat, water depth, duration, and temperature necessary for fish production, waterfowl breeding, and other uses;
- Support biodiversity (U.S. EPA 1993).

Grazing primarily affects four major components of the water-riparian system - banks/shores, the water column, channel, and aquatic and bordering vegetation.

Problems and Impacts

Stream crossing and pond bottom disruptions When livestock, particularly dairy cows and beef cattle, walk in streams and ponds, they disturb bottom composition, increase turbidity, and dislocate aquatic vegetation (Figure 2.11). Such disturbance can have serious impacts on the functioning of aquatic ecosystems.

Erosion and sedimentation Free-range livestock trample vegetation, compact dry soils, and loosen wet soils, thereby increasing the potential for soil erosion and runoff. In riparian areas, livestock disrupt stream morphology through bank destabilization and increased erosion.

Solutions

Structural

- Install fences around water bodies.
- Increase riparian vegetation and filter strips to mitigate impacts on water bodies.

Nonstructural

- Develop incentives to install fences to prevent cattle having access to water bodies.
- Create regulations to prevent use of riparian areas.

2.1.5.3 Crop Production

Description

Crop production often requires tillage and the application of fertilizers and pesticides to crops. Improper practices can have detrimental effects onsite and offsite in the watershed. After tillage, soil particles become susceptible to being displaced and transferred by wind and water erosion. Fertilizers and pesticides in runoff are nonpoint source pollutants that affect water resources, soils, and ecosystems.

Figure 2.12 Agricultural land with adequate riparian buffers

Problems and Impacts

Nutrient loading Overuse of fertilizers and failure to use conservation practices can cause nutrient contamination of surface waters. Groundwater resources, especially in sensitive areas that have high water tables, highly permeable soils, and high annual rainfall, can be affected by these pollutants. Animal manure that is used in crop fertilization can add high levels of nitrogen and phosphorus that can lead to water contamination. Drinking water supplies that are contaminated by nutrients pose health risks to both human beings and wildlife. Contaminated groundwater can be difficult to treat and expensive to restore. The relative levels of nitrogen and phosphorus are known to be a primary factor in excessive plant growth in aquatic systems, often leading to eutrophication.

Erosion Barren and tilled soils are prone to water and wind erosion. Vegetative cover with well-established roots can stabilize the soil. Water erosion is evident from water trails in the soil resulting from concentrated flows, gullies, washed off areas, and deposition and accumulation of soils in low-lying areas. Wind erosion is significant in dry and semiarid regions of the world. Both forces can displace significant amounts of watershed soils into water bodies and offsite locations.

Pesticides Insecticides and herbicides are commonly used in crop production to reduce crop losses. The type and amount of pesticide used varies according to crop type, soil conditions, and pest type. The use rate depends on the crop types in a region. For example, soybeans, cotton, corn, and wheat account for 70 percent of insecticides and 80 percent of herbicides applied to crops in the United States (Miller 2006). Pesticides are applied either directly on crops or added to soils. These chemicals can be washed off by runoff water or blown by wind into water resources. Pesticides can accumulate and persist in sediments and affect normal functioning of fish and other aquatic organisms. Pesticides can impact drinking water quality and public health.

Solutions

Structural

Nutrient management practices are effective in minimizing nutrient losses on farms. These include manure management, optimal fertilizer application, the choice of cropping systems, site management to minimize runoff, and implementation of nutrient conservation practices.

Erosion in croplands can be minimized through no-till or minimal tillage operations. Cropping systems that minimize soil erosion can be adopted after consideration of the slope conditions. Maintaining vegetative filter strips in sensitive areas can reduce sedimentation in water bodies (Figure 2.12).

Integrated pest management (IPM) can be used to manage pests in crop land. IPM programs use information on pests and their interaction with the environment, available pest control methods, economics, and impacts of infestations to manage pest populations.

Nonstructural

Incentive programs to encourage the adoption of conservation measures can be used. An example is cost sharing on material and structural components to encourage adoption of soil conservation methods.

Regulation of specific cropping practices that impact the watershed can be adopted in areas with serious problems.

Education and provision of timely information can encourage voluntary implementation of best management practices.

Case Study

Watershed Name: New York City water supply system
Location: Croton and Catskill/ Delaware Watershed System, New York, U.S.A.
Major Problem: Drinking water quality
Approaches: Collaboration between upstream farmers and the downstream city to link water quality objectives with an economic objective to preserve the watershed's farming economy.
Information:
http://www.usaid.gov/our_work/environment/water/case_studies/nyc.watershed.pdf

2.1.6 Forestry Practices

Forests provide recreational benefits, ecosystem services, and economic gains. A primary goal of forest management is to maintain a balance between the human needs for timber and other forest products and the habitat needs of forest ecosystems. Timber harvesting using heavy equipment can disturb soil and increase sedimentation in runoff (Figure 2.13). Many best management practices have been developed to minimize the effects of forest harvesting. Careful planning of forest operations is critical to avoid impacts on fragile ecosystems.

2.1.6.1 Road Construction and Skidder Trails

Description

Forest roads and skidder trails are required to transport timber from the logging areas to access areas for larger trucks that transport the product to the market. Often these roadways cross ecologically fragile forest land and can cause ecological stress to the connectivity of the ecosystem. Skidder trails and roads contribute up to 90 percent of the total sediment resulting from forestry operations (Rothwell 1983).

Problems and Impacts

Soil erosion Roads that are not properly designed and regularly maintained can erode quickly and load sediment into downstream waters.

Sedimentation is the deposition of eroded material from land surfaces into rivers and streams. Higher levels of sedimentation increase water turbidity, limit the ability of some aquatic organisms to survive, and can lower primary productivity of aquatic systems. Turbidity also reduces the transmission of light to photosynthetic organisms.

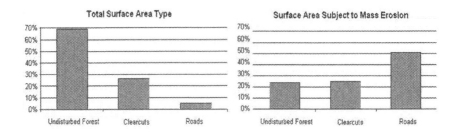

Figure 2.13 Contribution of roads to erosion in a forest ecosystem (Western Cascade Mountains, Oregon, U.S.A.). Data source: U.S. EPA (1993)

Solutions

Structural

BMPs can be designed to mitigate ecological impacts of roadways. Important factors to be considered include slope, stream crossings, and drainage patterns. BMPs can aid in increasing the service life of a road and minimize the impact of the road on the watershed system. Forest management plans should include erosion control measures that will be implemented. Forest plans can incorporate the following practices:

- Appropriately spaced water bars (ridges built across a trail) across skid trails.
- Vegetative filter strips in riparian areas.
- Drainage ditches and culverts of appropriate sizes in stream crossings and runoff areas.
- Properly placed BMPs and limited road use to reduce erosion.

Nonstructural

- Create regulations that use selective planning of logging areas with least impact on watershed ecosystems.
- Develop incentives to implement BMPs in forest operations.

2.1.6.2 Petroleum Spills

Description

Logging machinery and transport trucks introduce the risk of petroleum spills in a forest environment. Hydrocarbons such as petroleum and motor oil are primarily degraded in aerobic environments such as topsoil. As petroleum spills leach down into the water table, there is less oxygen available for degradation. The high carbon content in forest soils makes the hydrocarbons bond quickly, causing spills to be very persistent in soils and have poor potential for biodegradation.

Problems and Impacts

Spills Poorly maintained machinery and engines frequently leak oil from worn gaskets and loose tubing, which can cause small, but widely distributed petroleum pollution in the forest.

Loss of nutrients Hydrocarbons require a high amount of dissolved oxygen and nutrients to degrade. The chronic depletion of these nutrients can impact native organisms resulting in ecosystem changes.

Solutions

Structural

- Machinery maintenance is critical in order to identify potential leaks.
- Use protective covers for storing machinery and fuel, especially in areas of high risk.
- Respond rapidly to spills using surfactant-flushing techniques to flush out pollutants.

Nonstructural

- Require all machinery to be equipped with spill remediation equipment (such as absorptive sheets and degradation enzymes).
- Rigidly enforce vehicle maintenance codes to minimize chances of accidental spills.

2.1.6.3 Pesticides

Description

Silviculture operations often involve the use of pesticides to protect valuable forest stands. Herbicides are commonly used in forestry to prevent broad-leafed weeds, grasses, and hardwood shrubs from overtaking the profitable lumber trees. Sometimes insecticides are used to minimize the damage from leaf-eating insects and to prevent large-scale damage. Pesticides are used to clear areas of stumps and weeds. While the use of pesticides can increase stand productivity and timber yield, they can be harmful to non-target organisms. Most pesticides can persist in the environment for only a few weeks. Runoff can carry pesticides long distances from the source, causing damage before they are degraded.

Problems and Impacts

Non-target effects The death of non-target species such as grasses and moss causes changes in ecosystem functions. Pesticides can also destroy microorganisms and insects, disrupt the food chain, and decrease aquatic species biodiversity. These impacts have implications for the overall productivity of aquatic ecosystems.

Pesticide treadmill Organisms that are targeted with pesticides can eventually develop a genetic resistance to the toxins in the pesticide. Higher doses or new chemicals must be used to compensate for the loss in effectiveness, leading to higher costs of protection. This is often called the "pesticide treadmill".

Solutions

Structural

- Install drainage channels around stands to collect pesticides and bring them to a holding pond to allow them to biodegrade before entering the ecosystem.
- Use human power instead of pesticides to remove old stumps and weeds when preparing stands for planting.
- Use Drift-Free Pesticide Atomizer to reduce pesticide migration.
- Limit pesticide applications only to times when clear weather is forecast.
- Use pesticides only on stands that have a low chance of damaging fragile habitats.
- Use skidder or backpack pesticide applicators instead of applications from aircraft.
- In areas where much logging has occurred do not spray herbicides until enough foliage has resprouted.

Nonstructural

Research and utilize integrated and sustainable biological control mechanisms for pests such as natural predators or periodic flooding.

2.1.6.4 Forest Harvesting Techniques

Description

One of the major objectives in forest management is timber production. Recreation, habitat, game, and ecosystem protection are also important considerations in planning. Silvicultural techniques are used to maximize productivity and the quality of trees for wood products, with minimal impacts on the environment.

Problems and Impacts

The ecological damage from harvesting timber increases under the pressures from different harvesting methods. The following are examples of undesirable harvesting methods:

Clearcutting is the removal of the entire forest stand in one cutting. Seedlings are planted to revegetate the land. In this method, all trees are cleared but occasionally low value trees are left standing.

In the *seed-tree method*, mature timber is removed in one cutting. A small number of trees are left behind in small groups to provide seed for forest regeneration.

High-grading is a partial cut in which only the best trees are removed, leaving behind a residual stand of poorer quality. The objective is to achieve quick profits, but the practice could also lead to a loss of future forest productivity and quality of the ecosystem.

Solutions

Structural

Ecologically sound harvesting techniques can be adopted to increase the long-term productivity and health of a stand.

Salvage cutting is the removal of trees that are dead, dying or badly damaged before they decay. A *pre-salvage cutting* removes weak trees that are vulnerable to death or damage.

Shelterwood method Mature timber is harvested in a series of cuttings that extend over a relatively short time. This practice encourages establishment of even-aged reproduction.

Selection method Mature timber, usually the oldest or largest trees, is removed either as single, scattered individuals or in small groups at relatively short intervals. The process is repeated indefinitely.

2.1.6.5 Invasive Species

Description

Invasive species are non-native species that can cause economic and ecological disruptions to a watershed. These species are able to rapidly colonize and out-compete the native organisms that are essential for normal functioning of the watershed ecosystem. The uncontrollable proliferation of the invasive species can push native species to extinction, alter the habitat function, and impact the biodiversity of an ecosystem. Invasive species are recognized to be a significant problem for the sustainability of ecosystems. The most common biological invasions are the floras that result in gradual shifts in plant communities, which estranges wildlife and encourages further introduction of invaders.

Invasive plants, such as cheat grass, can increase fuel loads and create unpredictable fire hazards. Invasive weeds can decrease pine plantation yields by 63 percent (Westbrooks 1998). Plant invasions are very common because plant seeds can be distributed by boats, automobiles, or wildlife, thereby travelling over large areas within the germination season. Invasions also occur over longer distances by attaching to migratory species or through intense seasonal winds. Improper silviculture, land use changes, poor disposal of food, and high levels of ecological stress can enhance invasive plant growth.

Problems and Impacts

Habitat loss Changes in climax communities and the growing season can impact the habitat function of the ecosystem. These changes can result in high extinction rates, and a change of composition, structure, and function of the ecosystem.

Ecosystem dysfunction As the climax community shifts, native species are often the first to be pushed to extinction and replaced. The plant invaders can increase levels of erosion, increase the ecosystem's susceptibility to fire, flooding, and drought damage.

Solutions

Structural

- Roadside inspection stations on heavily travelled tourist routes can be used to find and quarantine invasive species.
- Airport monitoring stations can check for carry-on flora.
- Physical removal through hand weeding or tilling can destroy large invasions.
- Biocontrol methods can be employed where living organisms (insect herbivores or disease organisms) are introduced into populations of an invasive species.

Nonstructural

- Create regulation and inspection requirements for all boats, motors, and trailers.
- Educate conservationists and citizens about taxonomic identification.
- Provide adequate funding to local organizations to ensure rapid response to invasive species.
- Create new policy initiatives that provide incentives for voluntary adoption of protection methods in a watershed.

Case Study

Watershed Name: Tanzania coastal watersheds
Location: Tanzania, Africa
Major Problem: Mariculture and coastal zone management
Approaches: Integrated water resource management (IWRM) and sustainable mariculture development through partnerships.
Information:
http://www.usaid.gov/our_work/environment/water/case_studies/tanzania.mariculture.pdf

---

2.1.7 Recreation

Watersheds provide numerous opportunities for recreation such as boating, fishing, hunting, and hiking. Many recreational activities can alter habitats and resource conditions. For instance, the development of a golf course requires replacement of natural vegetation with turf and shrubs, creation of waterways, and sometimes the application of pesticides and fertilizers for maintenance.

Detrimental impacts on watershed systems can result from a variety of recreation activities. These include the building of marinas, petroleum spills, waste generation, trail erosion, depletion of water quality, habitat disturbance, and air pollution. It is important that recreational and watershed managers work together to enforce existing regulations, develop new strategies, and implement practices - including educating and securing the cooperation of recreational users - to minimize the potential negative effects of recreational activities on watersheds.

2.1.7.1 Marinas and Boating

Description

Boating is increasing in popularity throughout the world. As more people become involved in coastal recreation, habitats and resources can be at risk. The development of marinas is increasing significantly along inland waterways (Figure 2.14). Pollution from human usage and motorized boats has serious impacts on water quality and aquatic habitat. Elevated levels of metals and petroleum hydrocarbons can be toxic to aquatic organisms and other users. Inland rivers and lakes are also becoming increasingly stressed by people involved in non-motorized boating activities (kayaking, canoeing, and rafting), especially along put-in and takeout locations which are prone to bank instability and increased erosion (U.S. EPA 1993).

Problems and Impacts

Marinas located on the water's edge create concerns for environmental quality. Some problems result from a lack of vegetative buffers to filter nonpoint source pollutants. Environmental impacts in poorly flushed waterways include dissolved oxygen deficiencies, pollutants discharged from boats, and runoff from parking lots, roofs, and other impervious surfaces. These pollutants can also have serious impacts on wetlands and shellfish areas (U.S. EPA 1993).

Sewage and low dissolved oxygen In addition to being aesthetically unpleasant, untreated sewage discharged by recreational boats degrades water quality by adding harmful pathogens to water bodies. Water quality problems associated with human waste can force the closure of economically important shellfish beds and recreational sites in coastal waters. This decomposition of sewage depletes the total level of available oxygen in the water column. Poor flushing often exacerbates the problem (U.S. EPA 1993).

Figure 2.14 Riverside marina

Metals and metallic compounds are used in boat operation, maintenance, and repair. They are commonly found in the vicinity of these activities. Some metals include lead (fuels), arsenic (in paints, pesticides, and wood preservatives), zinc (used in prop anodes to deter corrosion of metal hulls and engine parts) and copper and tin (used as biocides in antifoulant paints). These metals settle into bottom sediments and are absorbed by many marine organisms, especially shellfish (U.S. EPA 1993).

Petroleum hydrocarbons The fuelling of motorized boats at marinas, pumping of bilges, and ruptured fuel lines can result in spills into water bodies. Many of these compounds, especially polynuclear aromatic hydrocarbons (PAHs), attach to organic matter and become incorporated into bottom sediments, especially under poor flushing conditions. Petroleum hydrocarbons degrade water quality and cause detrimental effects to aquatic life (U.S. EPA 1993).

Shoaling and shoreline erosion The creation of paths at put-in and takeout points along freshwater streams and lakes results in trampled vegetation and increased erosion and sediment loading in waterways. The effect on aquatic life is largely a result of the spatial disturbance of aquatic organisms and fringing plants (U.S. EPA 1993).

Introduced species Non-native and undesirable plant and aquatic species (e.g. milfoil and zebra mussels) can be introduced into waterways, transferred by boat hulls, motor props, and bilge and ballast waters (Figure 2.15). These introduced species cause serious problems for inland waterways and often become invasive, displacing native species, which can cause disruptions in food chains and shifts in the composition of ecosystems.

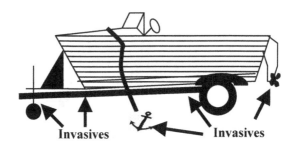

Figure 2.15 Invasive species may enter waterways on boats

Solutions

Structural

Installed Marine Sanitation Devices (MSDs), or heads, can be used to treat sewage. Types I and II MSDs utilize a chemical treatment which allows waste to be discharged into coastal waters, while type III MSDs are holding tanks which must be discharged more than three miles from shore. Vessels with MSD types I and II can still contaminate waters because the chemicals used to deodorize and disinfect sewage, which include alcohol, formaldehyde, zinc, and ammonium salts, degrade water quality (Massachusetts Bays Program 2003).

- Water depths in new marinas should support a natural population of benthic organisms.
- Fewer dock segments can achieve proper circulation within the basin.
- In poorly flushed water bodies an open marina, wave attenuation structures, and the proper location of entrance channels can be used to enhance flushing. Two openings at opposite ends of the marina can promote flow-through currents.
- Implement erosion control practices in the marina.

Nonstructural

- Create regulations to require cleaning of all plant material attached to motor props and boat hulls before launching vessels. Require that bilge waters be pumped before boats leave waterways.

- Educate boaters through signs at launch points to alert them to the problem of invasive species and prevention methods.
- Use grant programs to encourage conservation practices in construction, renovation, operation, and maintenance of boat pump-out and waste reception facilities at public and private marinas (Massachusetts Bays Program 2003).

Case Study

Watershed Name: Tana River Basin
Location: Southeastern Ethiopia, Africa
Major Problem: Poverty and access to safe water
Approaches: Economic recovery strategy involves communities and local authorities in management of water and sewer systems and services.
Information:
http://www.unesco.org/water/wwap/wwdr2/case_studies/pdf/kenya.pdf

---

2.1.7.2 Trail Use

Description

The use of outdoor trails for recreation such as hiking, biking, wildlife viewing, and horseback riding is gaining popularity. Improper trail development and use can have negative environmental impacts, either from habitat alteration, littering, increased erosion, and soil compaction.

Problems and Impacts

Erosion Heavily travelled trails are prone to increased erosion from foot traffic, horseback riding, pack animals, and mountain bikes, which can displace soil and disturb soil substrate. These soils are vulnerable to runoff, leading to increased sedimentation in downstream waterways and the creation of gullies along trails.

Solid waste Besides being aesthetically displeasing, trash left behind by people using recreational trails can attract wildlife and nuisance animals, and contaminate water bodies with pathogens and harmful bacteria.

Nutrient enrichment Fecal matter and urine from pack animals such as horses and llamas, domestic dogs, and humans can introduce harmful bacteria and pathogens into water bodies.

Solutions

Structural

- Minimize use of horses and mountain bikes (seasonal or permanent trail restrictions) in environmentally sensitive areas.
- Design and construct new trails carefully, particularly in sensitive areas of the watershed.

Nonstructural

- *Public education* The majority of people who use trails for recreation may be unaware of good trail use practices. Making this information available through signs posted at trail entrances or other locations is an important tool to minimize trail use impacts. Patrolling of trails by rangers, who can educate the public and enforce regulations, is also useful.
- Entry fees and user limits can be used to regulate trail use in high demand areas.

SUGGESTED READING

2.1.1 – 2.1.2 Land Use Impacts on Watersheds, Residential and Municipal

Baehr, A. and Corapcioglu, M. Yavuz (1984) A Predictive Model for Pollution From Gasoline In Soils and Groundwater. In *Petroleum Hydrocarbons and Organic Compounds in Groundwater - Prevention Detection, and Restoration*, pp. 144-155, National Well Water Association, Worthington, Ohio.

Brooks, Kenneth P., Ffolliott, H., Gregersen, H. and DeBano, L.F. (2003) *Hydrology and the Management of Watersheds,* 3rd edn, Iowa State Press, Ames, Iowa.

Brown, K.W. and Donnelly, K.C. (1988) An estimation of risk associated with the organic constituents of hazardous and municipal waste landfill leachates. *Haz Waste Haz Matter.* **5**(1), 1-30.

Center for Watershed Protection (1998) *Rapid Watershed Planning Handbook: A Comprehensive Guide For Managing Urbanizing Watersheds*, Center for Watershed Protection, Inc., Elliott City, Maryland.

Massachusetts Bays Program (1998) *Massachusetts Bays Watershed Stewardship Guide: An Education Resource.* URL: http://www.msp.umb.edu/mbea/mbeaguid.htm.

MA DEP (1993) *Landfill Technical Guidance Manual.* MA Department of Environmental Protection, Boston, Massachusetts.

MA DEP (2006) *Massachusetts Nonpoint Source Pollution Management Manual.* URL: http://projects.geosyntec.com/megamanual/.

National Research Council (1995) *Mexico City's Water Supply: Improving the Outlook for Sustainability*, Academia Nacional de Ingeniería, National Academy Press, Washington, D.C.

Ostendorf, D.W., Noss, R.R., and Lederer, D.O. (1984) Landfill Leachate Migration through Shallow Unconfined Aquifers. *Water Resources Research.* **20**, 291-296.

U.S. EPA (1993) *Guidance specifying management measures for sources of nonpoint pollution in coastal waters.* Report EPA 840-B-92-002, U.S. Environmental Protection Agency, Office of Water, Washington, DC.

U.S EPA (1994) *Developing Successful Runoff Control Programs For Urbanized Areas.* Report EPA 841-K-94-003, U.S. Environmental Protection Agency, Office of Water, Washington, D.C.

U.S. EPA (1996a) *Environmental Indicators of Water Quality in the United States.* Report EPA 841-R-96-002, U.S. Environmental Protection Agency, Washington, D.C.

U.S. EPA (1996b) *Municipal Wastewater Management Fact Sheets: Stormwater Best Management Practices.* Report EPA 832-F-96-001, U.S. Environmental Protection Agency, Washington, D.C.

U.S. EPA (1998) *Clean Water Action Plan: Restoring and Protecting America's Waters.* Report EPA 840-R-98-001, U.S. Environmental Protection Agency, Office of Water, Washington, D.C.

U.S. EPA (2000) *Low impact development: A literature review.* Report EPA 841-B-00-005, Office of Water, U.S. Environmental Protection Agency, Washington, DC.

U.S. EPA (2001) *The Brownfields Economic Redevelopment Initiative, Proposal Guidelines for Brownfields Cleanup.* Report EPA 500-F-01-348, Solid Waste and Emergency Response, Washington, D.C. URL: http://72.14.207.104/search?q=cache:_gYymWlOLxMJ:www.epa.gov/brownfields/pdf/bcrlfgui.pdf+Environmental+Protection+Agency.+1996.+The+Brownfield+Economic+Redevelopment+Initiative%3B+Application+Guidelines+for+Brownfields+Assessment+Demonstration+Pilots.&hl=en&gl=us&ct=clnk&cd=2&client=firefox-a.

U.S. EPA (2002) *Small Business Liability Relief and Brownfields Revitalization Act.* URL: http://www.epa.gov/swerosps/bf/sblrbra.htm.

U.S. EPA (2006d accessed) *How to Conserve Water and Use It Effectively.* U.S. Environmental Protection Agency, Office of Water. URL: http://www.epa.gov/OW/you/chap3.html.

U.S. EPA (2006f accessed) *Superfund, Cleaning Up the Nation's Hazardous Waste Sites.* URL: http://www.epa.gov/superfund/.

Winkler, E.S. (1998) *Innovative and Alternative On-Site Wastewater Treatment Technologies Handbook.* UMASS Extension Bulletin Center, University of Massachusetts Amherst. Amherst, Massachusetts.

2.1.3 – 2.1.4 Construction and Mining

Franklin, Hampden, Hampshire Conservation Districts (1997) *Massachusetts Erosion and Sediment Control Guidelines for Urban and Suburban Areas: A Guide for Planners, Designers and Municipal Officials.* Prepared for Massachusetts Department of Environmental Protection, Executive Office of Environmental Affairs, Northampton, Massachusetts.

Massachusetts Bays Program (1998) *Massachusetts Bays Watershed Stewardship Guide: An Education Resource.* URL: http://www.msp.umb.edu/mbea/mbeaguid.htm.

Miller, G. Tyler, Jr. (2006) *Environmental Science*, 11th edn, Wadsworth Publishing Co., Belmont, California.

2.1.5 – 2.1.7 Agriculture, Forestry and Recreation

Agrios, George N. (1997) *Plant Pathology*, 4th edn, Academic Press, London, UK.

Kittredge, David B. Jr. and Parker, Michael (1999) *Massachusetts Forestry Best Management Practices Manual*. MA Department of Environmental Protection, Boston, Massachusetts.

Massachusetts Bays Program (2003) *Massachusetts Bays Comprehensive Conservation & Management Program: An Evolving Plan For Action*. URL: http://www.mass.gov/envir/massbays/pdf/revisedccmp.pdf.

Miller, G. Tyler, Jr. (2006) *Environmental Science*, 11th edn, Wadsworth Publishing Co., Belmont, California.

Oregon State University (2006) *Integrated Plant Protection Center*. URL: http://www.ipmnet.org/IPM_Handbooks.htm

Radcliffe, E.B. and Hutchison, W.D. (2006) (eds.), *Radcliffe's IPM World Textbook*, University of Minnesota, St. Paul, MN. URL: http://ipmworld.umn.edu.

Rothwell, R.L. (1983) Erosion and sediment production at road-stream crossings. *Forestry Chronicle*. **23**, 62-66.

Ryan, David K., Duggan, John W. and Bruell, Clifford J. (1995) *Enhanced Recovery of Gasoline Hydrocarbons by Soil Flushing with Solutions of Dissolved Organic Matter and Nonionic Surfactants*. Report 170, Water Resources Research Center, University of Massachusetts, Amherst, Massachusetts.

U.S. Department of Agriculture (USDA) (1994) *Evaluating the Effectiveness of Forestry Best Management Practices in Meeting Water Quality Goals or Standards*. Report 1520, USDA, U.S. Forest Service, Southern Region, 1720 Peachtree Road NW, Atlanta, Georgia 30367.

USDA (1998a) *National Extension Targeted Water Quality Program, 1992-1995, Outcomes of Animal Waste Programs*. vol. 2, U.S. Department of Agriculture, Cooperative Extension Service, Washington, D.C.

USDA (1998b) *National Extension Targeted Water Quality Program, 1992-1995, Outcomes of Crop Pesticide Management Programs*. vol.4, U.S. Department of Agriculture, Cooperative Extension Service, Washington, D.C.

USDA (1998c) *National Extension Targeted Water Quality Program, 1992-1995, Outcomes of Nitrogen Fertilizer Management Programs*. vol.3, U.S. Department of Agriculture, Cooperative Extension Service, Washington, D.C.

U.S. EPA (1993) *Guidance Specifying Management Measures For Sources of Non-point Pollution in Coastal Waters*. Report EPA-840-B-92-002, U.S. Environmental Protection Agency, Washington, D.C.

U.S. EPA (1998) *Clean Water Action Plan: Restoring and Protecting America's Waters*. Report EPA 840-R-98-001, U. S. Environmental Protection Agency, Office of Water, Washington, D.C.

Westbrooks, R. (1998) *Invasive plants, changing the landscape of America: Fact book*. The Federal Interagency Committee for the Management of Noxious and Exotic Weeds (FICMNEW), Washington, D.C.

Worrall, Jim (2005) *Selected References in Forest Pathology*. Forest and Shade Tree Pathology. URL: http://www.forestpathology.org/refs.html.

3

Inland Water Bodies

The rivers, streams, lakes, groundwater and wetlands that comprise inland water bodies are vital resources for human beings, both for their water supplies and their ecosystem services. The diverse plant and animal life associated with inland water bodies are important sources of livelihood to people all over the globe. Consequently, these ecosystems have been significantly altered by human beings and are at risk due to over-exploitation of water supplies and aquatic life, extensive pollution, loss of biodiversity and the introduction of invasive species. According to the United Nations Convention on Biological Diversity, 41 percent of the world's population live in river basins that are water stressed. Approximately 20 percent of inland fish species are threatened or extinct, a much larger amount than for marine species (UNEP 2005a). The management of inland waters is increasingly recognized as vital to sustainable development.

In the United States since the passage and implementation of the Clean Water Act of 1972, America's inland water resources have experienced significant improvements in water quality. Restoration of aquatic ecosystems, enhanced protection of aquatic plants and wildlife, increased recreational opportunities and economic benefits have followed. In 1972, it was estimated that only 30 to 40 percent of waters met quality standards for fishing and swimming. Today, reports from state monitoring programs nationwide indicate a compliance rate of 60 to 70 percent (U.S. EPA 1998).

However, serious water pollution problems persist in inland waterways across the United States and around the world. The degradation of water quality by nonpoint source pollution continues to threaten drinking water supplies, public health, and the health and functioning of aquatic ecosystems.

3.1 RIVERS AND STREAMS

Rivers and streams provide many important benefits to society, including recreational values such as fishing and boating, economic values such as power generation and sewage treatment disposal, and environmental values such as drinking water and habitat for aquatic species. Because rivers and streams are moving systems that can cover vast expanses of land, the negative effects of land use activities that lead to the degradation of these waterways in turn affect the watersheds through which they flow. Nonpoint source pollution from runoff water is a primary cause of water quality problems in rivers and streams, but a number of point sources can degrade rivers and streams as well.

3.1.1 Riparian Systems

Description

Riparian areas are distinct transitional zones between terrestrial and aquatic ecosystems. Due to their boundary niche, riparian zones exhibit sharp environmental gradients. Soil moisture content in riparian areas is in excess of that normally available due to runoff and/or subsurface seepage. This results in an existing or potential soil-vegetation complex characteristic of very moist soils (Brooks et al. 2003). When left in their natural state, or when well managed, riparian systems perform valuable ecological services. They filter upstream runoff water, provide diverse and dynamic habitat, influence riverine morphology, mitigate flood impacts, and provide economic services such as recreation. Riparian areas may be concentrated in a small geographical location, but their value extends far beyond into the watershed system. The alteration or destruction of riparian areas can result in serious consequences for the chemical, physical, and biological integrity of inland water bodies and watershed systems. Most of these effects are directly or indirectly related to changes in water quality, habitat, and economic activities.

Problems and Impacts

Vegetation removal Removal of trees and vegetative ground cover from these sensitive areas can have serious impacts on water quality. Streamside vegetation maintains cooler water temperatures by reducing the amount of solar radiation reaching the stream surface. Cooler temperatures support certain sensitive aquatic organisms and maintain healthy levels of dissolved oxygen. Vegetation also stabilizes the banks of streams, rivers, ponds, and lakes, thereby reducing the amount of bank cutting and erosion, and decreasing sediment loading into the water body. In addition, vegetation filters runoff from upland area, thereby decreasing the levels of sediment, nutrient and other contaminants in the water.

Bank collapse Root systems of riparian vegetation bind the soils, providing stability to stream banks. The removal of riparian trees and their root systems can result in the instability of the soils and banks of rivers and streams. The impact is readily visible as gully formations, eroded and collapsed banks, and toppled trees in waterways. The addition of large woody debris, such as toppled trees, may provide hydrologic and ecological benefits to streams and rivers. However, increased erosion and sedimentation resulting from frequent bank collapse are detrimental to aquatic ecosystems.

Erosion When the banks of streams, rivers, ponds, and lakes are made unstable by vegetation removal, they become susceptible to erosive forces of wind and water. Stormwater runoff from upland areas can add excess, unwanted sediment to the system.

Sedimentation This results from erosion and the introduction of soil particles into rivers and streams. Sedimentation causes turbidity, buries fish and insect habitat and their eggs, and causes physical changes in stream channels.

Water temperatures A significant rise in water temperature can result from increased solar radiation in riparian areas with poor vegetative cover. Increases of a few degrees in water temperature can seriously affect the ecological balance of rivers and streams because of impacts on the health and viability of specific species.

Nutrient enrichment Poor riparian systems cannot provide the function of nutrient uptake and filtering. This results in nutrient loads from agriculture and development activities directly reaching water bodies.

Dissolved oxygen (DO) can be depleted by biological and chemical activity that occurs as a result of nutrient enrichment and increased water temperature. Environments with decreased DO are not suitable habitat for fish and aquatic plants.

Infiltration Riparian trees and vegetative ground cover facilitate infiltration of water into soils. Removing vegetation inhibits infiltration and disrupts the water balance of riparian systems.

Food chain disruption All of the above-mentioned impacts have implications for the aquatic food chain, individually as well as cumulatively. The removal of trees along streams can cause a significant decrease in the amount of litter fall such as leaves, twigs, and fruit entering the stream. Reduced litter fall decreases the productivity of detritus-eating macroinvertebrates, which in turn can affect fish and other vertebrate populations.

Case Study

Watershed Name: Rhone-Mediterranean Basin
Location: Southeastern France and Switzerland, Europe
Major Problem: Dam impact on instream flows
Approaches: Action plans to increase water flows.
Information:
http://unesdoc.unesco.org/images/0014/001459/145925E.pdf

3.1.2 Large Woody Debris and Large Organic Debris

Description

Large woody debris (LWD) and large organic debris (LOD) are important components of river and watershed systems. LOD is any relatively stable, woody material that has a diameter greater than 10cm and a length greater than 1m that intrudes into the stream channel (Brooks et al. 2003). LWD specifically refers to root wads and tree stems that provide overhead cover and flow modification for effective spawning and rearing habitat for anadromous and resident fishes (American Fisheries Society 1985). LWD plays an important role in hydraulics, sediment routing, channel morphology, stream complexity and the formation of suitable habitat for aquatic organisms (Brooks et al. 2003).

Problems and Impacts

Channel morphology Large, affixed logs lying across the channel deflect the current laterally, which can result in widening of the streambed and erosion of the shoreline. Scour pools form below obstructions, deepening the channel, while sediment stored by debris adds to hydraulic complexity. LWD contributes to habitat diversity in streams by anchoring and stabilizing the position of the pools in the direction of the stream flow. LWD helps to create backwaters along the stream margin, causing a lateral migration of the channel and forming secondary channel systems in alluvial valley floors. The debris also increases depth variability (Brooks et al. 2003).

Hydraulics LWD offers hydraulic resistance in forest streams, dissipating energy and changing erosive pressures (Brooks et al. 2003).

Sediment routing Sediment, which is trapped by LWD and LOD, adds to the hydraulic complexity and can create well-developed secondary channel systems (Brooks et al. 2003).

Habitat Organic debris in streams can increase aquatic habitat diversity. LOD aids in the formation of pools and protected backwater areas, provides nutrients and substrate for biological activity, dissipates the energy of flowing water, and traps sediment (Brooks et al. 2003).

3.1.3 Invasive Species

Description

Invasive species can cause disruptions in ecosystem function and alter the physical characteristics of water bodies, such as flow rate in waterways. Invasive plants are common in waterways because they can be easily transported as small seeds and can survive outside of water for extended periods of time. Many inland waterways are so overrun by invasive species that management efforts are focused on control rather than eradication of the invaders.

Problems and Impacts

Aquatic competition Invasive plants with broad leaves can reduce light transmittance through the water column, limiting the energy required for photosynthetic organisms. Species that proliferate early in the season block light for plants with slower growth and limit the native plant productivity.

Channel alteration Over time, as invasive species dominate the waterway, detritus from decaying native species accumulates on the bottom. In some areas plants have been able to fill in coves and alter the sinuosity of rivers with massive collections of plant debris and sediment.

Solutions

Structural

Mechanical/manual harvesting An easy method of eradicating unwanted species is through manual removal; however, it requires large amounts of money and human resources to accomplish the task. Mechanical harvesters can cut aquatic weeds to deep below the water surface and usually carry between a half and two tons of weeds until they can be disposed of properly. Mechanical harvesting is often used for plant removal in heavily infested areas.

Chemical sprays Application of pesticides is a cost-effective method for controlling some invasive plants. Plant-specific and easily degradable pesticides are continually being developed. Government organizations are often reluctant to use chemicals in an attempt to set a positive example and avoid any potential adverse impacts to aquatic life.

Sustainable management The control of invasive species using natural predators and parasites is a preferred method when it can be done without risk of introducing new species or a new disease. This method requires both extensive testing, knowledge of the pest and predator, and often takes a longer time to eradicate the invader completely.

3.2 LAKES AND PONDS

3.2.1 Eutrophication

Description

When a body of water receives nutrients such as nitrogen and phosphorus in excess of that which the system can naturally handle, the introduced nutrients can accelerate microscopic algae (plant) growth. When algae die, they accumulate in the water column and during the process of decomposition, they deplete the dissolved oxygen in the water. Anoxic (no dissolved oxygen) or hypoxic (low dissolved oxygen) conditions often lead to the death of fish and other aquatic organisms. This process is called eutrophication. Eutrophication is a common problem near agricultural and urban areas. Urban activities such as construction and landscaping increase erosion and nutrient loads in runoff that are discharged into water bodies. In agricultural and residential areas fertilizers can enter runoff and accumulate in water systems. Septic systems can leak nutrients into groundwater and water bodies (see also Chapter 6).

Problems and Impacts

Excessive plant growth In lakes and ponds, symptoms of eutrophication include excessive aquatic plant growth, low levels of dissolved oxygen, a change in water color to green, increased turbidity, and odors from fish and plant decomposition.

Fish kills The deaths of large numbers of fish can be a result of low dissolved oxygen triggered by decomposition of plants and the die-off of a large algal bloom. Large fish kills can result in a shift in species composition in aquatic ecosystems.

Socioeconomic impacts Eutrophication changes the ecology of a water body, with a consequent loss of services such as recreational use for swimmers and boaters, decreased property values for waterfront homes, low fish harvests, and increased costs for treating drinking water.

Solutions

Structural

Protect and create vegetative buffer strips Vegetative buffer strips can filter nutrients before they reach water bodies. Where vegetation exists, leave a sufficient width of riparian buffer along lakeshores, riverbanks, and intermittent streams. Where there is no riparian vegetation, buffer strips can be created by planting deep-rooted, woody vegetation that will stabilize the shorefront, prevent erosion, and filter nutrients.

Erosion control Land owners and users need to minimize land disturbance using erosion control and nutrient management practices. These include fabric filter fences in construction areas, hay bale dams to control erosion, and appropriate cropping practices. Land owners should also direct ditches away from water bodies and into vegetated areas, mulch and replant exposed soils within one week, and avoid construction on high slopes areas.

Leach fields and septic systems Septic systems should be kept in good working condition, and leach fields should be kept at a reasonable distance (depending on local characteristics) from lake shores.

Nonstructural

Conservation incentives can be designed to encourage upstream watershed users to implement best management practices.

Regulations can be implemented using standards based on eutrophication potential.

Trading of pollutant permits (markets) can be established at a watershed level to minimize contaminants to the water body.

3.3 WETLANDS AND VERNAL POOLS

Freshwater marshes, swamps, bogs, saltwater marshes and tidal flats are examples of wetlands. According to the U.S. Fish and Wildlife Service, wetlands are transitional areas between terrestrial and aquatic systems. For the purpose of this classification wetlands must have one or more of the following attributes: 1) at least periodically, the land supports predominately hydrophytes; 2) the substrate is predominately undrained hydric soil; and 3) the substrate is nonsoil and is saturated with water or covered by shallow water at some time during the growing season of each year (Cowardin et al. 1979).

Worldwide, wetlands cover approximately 7 to 9 million km^2 or roughly 4 to 6 percent of the earth's land surface (Mitsch and Gosselink 2000). They are among the most productive of ecosystems and are vital resources due to the various ecological roles they play.

Depending on their type and location, wetlands improve water quality, facilitate groundwater recharge, stabilize shorelines, reduce floodwaters and trap sediment (Figure 3.1). Wetlands also provide crucial fish and wildlife habitat for 5,000 species of plants, 190 species of amphibians, and a third of all the bird species in the United States (USDA 1995).

Fish and wildlife values	Environmental quality values	Socioeconomic values
• Fish and shellfish habitat	• Water quality maintenance	• Flood control
• Waterfowl and other bird habitat	• Pollution filter	• Wave damage protection
• Furbearer and other wildlife habitat	• Sediment removal	• Erosion control
	• Oxygen production	• Groundwater recharge and water supply
	• Nutrient recycling	• Timber and other natural products
	• Chemical and nutrient absorption	• Energy source (peat)
	• Aquatic productivity	• Livestock grazing
	• Microclimate regulator	Fishing and shellfishery
	• World climate (ozone layer)	• Hunting and trapping
		• Recreation
		• Aesthetics

Figure 3.1 Wetland values (Tiner, 1984)

According to the U.S. Fish and Wildlife Service (1984), of the roughly 99 million acres of wetlands remaining in the lower 48 states, 94 percent are inland freshwater wetlands, more than half of which are forested. The remaining 6 percent are coastal wetlands. Despite the growing awareness of the important environmental and economic value of wetlands, they continue to be destroyed at a rate of 450,000 acres per year. As a result, less than 46 million of the original 215 million acres of wetlands in the lower 48 states remain today.

Vernal pools, also known as spring ponds, ephemeral wetlands, coastal bays, or whale wallows, are temporary wetlands, which fill with water sometime between fall and spring and are usually dry by late summer. Vernal pools are essential for the life cycle of many invertebrates and amphibians.

3.3.1 Drainage, Filling, and Alteration

Description

Wetlands serve as natural flood control systems by storing stormwater. The ability of wetlands to perform this function can be destroyed by draining and filling for irrigation and urban development. Serious impairment of wetland function may occur as a result of the alteration of a wetland's structure during construction of dams, ports and harbours, channelization projects, and shoreline stabilization.

Problems and Impacts

Functional loss occurs when wetlands lose their ability to perform their normal functions of water retention, storage, and purification.

Waste dumping Wetlands have been used as dumping grounds for decades because of the false belief that wetlands have no ecological value. Wetlands have only recently been recognized as vital resources and been protected from activities such as the disposal of solid, toxic, and hazardous wastes. Now that public officials understand the important role that wetlands play in protecting water quality and providing habitat for plants and wildlife, regulations and public education programs are in place or being developed to address this serious issue.

Solutions

Nonstructural

Wetland mitigation banking is used to maintain or increase the area of wetlands.

Regulation is a viable approach to protect wetlands. In the U.S. a number of federal laws and programs directly and indirectly provide some protection for wetlands (Barton 1986):

- Section 404 of the Clean Water Act requires issuance of a permit from the U.S. Army Corps of Engineers for the discharge of dredged or fill material into waters of the United States.
- The National Environmental Policy Act (42 U.S.C.A. 4321-4361) requires federal agencies to consider the environmental impacts of their activities, including those affecting wetlands.
- The Coastal Zone Management Act (16 U.S.C.A. 1451-1464) encourages and funds states to develop coastal management plans that include wetland protection measures.
- The Coastal Barrier Resources Act (16 U.S.C.A. 3501 *et seq.*) eliminates federal insurance and subsidies for development on certain barrier islands.
- The Executive Orders on Floodplain Management (E.O. 11988) and Protection of Wetlands (E.O. 11990) require federal agencies to take action to eliminate wetland destruction and to preserve their beneficial values.
- Federal land protection systems, including the Wild and Scenic Rivers System, the National Wilderness Preservation System, and the National Park System, prevent development of the wetlands included within their boundaries.

Case Study

Watershed Name: Danube River Basin
Location: Covers all or parts of Albania, Austria, Bosnia-Herzegovina, Bulgaria, Croatia, the Czech Republic, Germany, Hungary, Italy, the Former Yugoslav Republic of Macedonia, Moldova, Poland, Romania, Serbia and Montenegro, the Slovak Republic, Slovenia, Switzerland and Ukraine
Major Problem: Sharing water and water management
Approaches: International commission formed to coordinate sustainable water management practices involving water quality, flood control, and emission standards.
Information:
http://www.unesco.org/water/wwap/wwdr2/case_studies/pdf/danube.pdf

SUGGESTED READING

3.1 Rivers and Streams

American Fisheries Society (1985) *Aquatic habitat Inventory: Glossary and Standards Methods*, (ed W.T. Helm), pp. 1-34, Habitat Inventory Committee, Washington, D.C.

Barbour, M.T., Gerritsen, J., Snyder, B.D. and Stribling, J.B. (1999) *Rapid Bioassessment Protocols for Use in Streams and Wadeable Rivers: Periphyton, Macroinvertebrates and Fish, Second Edition.* Report EPA 841-B-99-002, U.S. Environmental Protection Agency, Office of Water, Washington, D.C. URL: http://www.epa.gov/owow/monitoring/rbp/

Brooks, Kenneth P., Ffolliott, H., Gregersen, H. and DeBano, L.F. (2003) *Hydrology and the Management of Watersheds*, Iowa State Press, Ames, Iowa.

UNEP (2005a) *Inland Water Biodiversity*. Secretariat of the Convention on Biological Biodiversity, United Nations Environmental Program, UNEP-CBD, Montreal, Canada.

U.S. EPA (2000) *Stressor Identification Guidance Document*. Report EPA-822-B-00-025, U.S. Environmental Protection Agency, Office of Water, Washington, D.C. URL: http://www.epa.gov/waterscience/biocriteria/stressors/stressorid.pdf.

U.S. EPA (2006h accessed) *Useful Links to Invasive Species Information*. URL: http://www.epa.gov/owow/invasive_species/links.html.

3.2 – 3.3 Lakes, Ponds, Wetlands and Vernal Pools

Barton, K. (1986) *Federal Wetlands Protection Programs*, Audubon Wildlife Report, National Audubon Society, New York.

Cowardin, L.M., Carter, V. F., Golet, C. and Laroe, E.T. (1979) *Classification of Wetlands and Deepwater Habitats of the United States.* Report FWS/OBS-79/32, U.S. Department of Interior, Fish and Wildlife Service, Washington, D.C.

London Biodiversity Partnership (2006) *Lakes, ponds and habitat audit*. URL: http://www.lbp.org.uk/02audit_pages/au13_ponds.html.

Mitsch, William J. and Gosselink, James G. (2000) *Wetlands*, 3rd edn, John Wiley & Sons, New York.

Tiner, Ralph Jr. (1984) *Wetlands of the United States: Current status and recent trends*, U.S. Department of Interior, Fish and Wildlife Service, Washington, D.C.

USDA (1995) *Wetland Values and Trends*. RCA Issue Brief #4, U.S. Department of Agriculture, Natural Resources Conservation Service, Washington, D.C.

U.S. EPA (1991) *Volunteer Lake Monitoring*. Report EPA 440-4-91-002, U.S. Environmental Protection Agency, Office of Water, Washington, D.C. URL: http://www.epa.gov/volunteer/lake/lakevolman.pdf.

U.S. EPA (2006e accessed) *Lake and Reservoir Bioassessment and Biocriteria, Technical Guidance Document.* U.S. Environmental Protection Agency, Office of Water, Washington, D.C.

URL: http://www.epa.gov/owow/monitoring/ tech/lakes.html.

4

Coastal Watersheds

Coastal watersheds have geographic areas that are influenced by both freshwater and seawater. The diverse habitats of coastal watersheds provide a wide variety of coastal and marine organisms with food, shelter, and breeding and nursery grounds. With increasing numbers of people moving to coastal areas, the development and pollution that accompany this population pressure are difficult to manage. Furthermore, coastal watersheds are influenced not only by the surrounding activities and land uses, but also by upstream land uses. Pollutants and sediment that may enter a river in an upland area will be carried downstream to coastal systems. The unique location of coastal watersheds makes them particularly vulnerable to habitat degradation and water quality problems.

An important characteristic of coastal watersheds is the presence of coastal ocean and estuarine zones. These are areas that interact with the ocean and include diverse ecosystems such as salt marshes, coastal and inter-tidal areas, bays, harbors, lagoons, inshore waters, and channels. Estuaries are defined as "all or part of the mouth of a navigable or interstate river or stream or other body of water having unimpaired natural connection with open sea and within which the sea water is measurably diluted with freshwater derived from land drainage" (National Estuary Study 1970).

Estuaries are ecologically productive areas, largely owing to the influx of enriching land-derived materials that are distributed via currents, tides, and wave action. It is this richness that is responsible for the biological diversity and ecological productivity making estuaries a vital component of the overall health of coastal ecosystems. When pollutants enter estuaries and coastal ecosystems, they can have serious effects on coastal habitats and on aquatic plants, invertebrates, mollusks, and other species of marine life. Changes in water temperature, oxygen levels, contaminant loads, and water levels caused by human activities can also negatively impact marine organisms and water quality.

4.1 HABITAT ALTERATION

Description

Habitat alteration is a common problem where human interaction with coastal areas is high. New roads, power lines, water diversions, tourism, waste generation, and residential and commercial development all have very serious and long-lasting consequences for wildlife.

Problems and Impacts

Habitat loss Marshes, tidal flats, and buffer zones are often the first ecosystems to be destroyed by development impacts. Serious disruptions can result in mass dislocations or local extinctions of wildlife.

Declining anadromous fish Anadromous fish migrations have been impaired in many parts of the world. For example, migrations have been declining along most of the U.S. coastal waterways for the past 100 years. This decline is due to pollution of spawning grounds and pathways, contamination of sediments, eutrophication of water bodies, and channel impoundments. Obstructions such as dams prevent the passage of adults and juveniles between freshwater breeding areas and the sea.

Declining eelgrass beds Eelgrass beds serve several critical functions in a coastal environment. They provide habitat for many species of finfish, shellfish, and waterfowl; reduce turbidity, and improve water quality by filtering suspended sediments and serving as a baffle to moving sand. Eelgrass beds also function as an essential component of near shore food webs, and provide nursery and feeding grounds for a number of commercially and ecologically important fish species. These beds are sensitive to and are threatened by numerous pollution sources, including sewer and stormwater discharges, dredge and fill activities, heavy boat traffic, and nonpoint pollution sources such as septic systems, agriculture, and urban runoff.

Solutions

Structural

- *Fish ladders* Installation and maintenance of fish ladders and passages are common and effective methods for reconnecting migration pathways.
- Restoration of coastal ecosystems through comprehensive methods at a watershed scale.

Nonstructural

- Zoning regulations and policies can create protected zones to safeguard vulnerable areas.
- Coastal zone management plans can be developed to protect sensitive habitat.

Case Study

Watershed Name: Mississippi River Watershed
Location: Central region of the U.S.A.
Major Problem: Nutrient contamination
Approaches: Integrated assessment and action planning through consensus.
Information:
http://www.epa.gov/msbasin/index.htm
http://pubs.usgs.gov/circ/circ1133/nutrients.html

4.2 COASTAL SEDIMENT TRANSPORT AND SHORELINE STABILITY

Description

Naturally occurring meteorological, hydrologic, and oceanographic forces affect coastal shorelines through sediment transport and hydrologic action. The transport of sediment in near shore waters results from complex interactions between waves, currents, sediments, and bed roughness. Sediment transport involves the interaction of flow and an erodible bottom caused by fluid motion at all frequencies. Erosion of shorelines can be caused by both long-term sediment transport and quick catastrophic change from storm events. The resulting increased shoreline instability can destroy waterfront real estate and hinder the national economy by creating navigational hazards to coastal shipping. Human activity contributes to shifting shorelines as increasing demand for coastal resources and shoreline development bring with it a wealth of problems. Human activities, which affect climate, are also exacerbating shoreline stability. For example, global warming and the rise in sea level may be compounding the problems of changing shorelines.

4.3 SHALLOW EMBAYMENTS

Problems and Impacts

Poor flushing, eutrophication Shallow bays and estuaries are prone to eutrophication when poor flushing conditions are coupled with incoming sources of nitrogen (Figure 4.1). Nitrogen introduced into these areas may come from ocean water inflow, sewage outfalls, upstream discharges, groundwater flow, atmospheric deposition, and stormwater runoff. The impacts of eutrophication on habitats lead to excessive algal and other aquatic plant growth resulting in decreased oxygen levels and limited sunlight penetration. The consequent disruption of photosynthetic processes, and decrease in oxygen levels can cause suffocation of fish and other fauna.

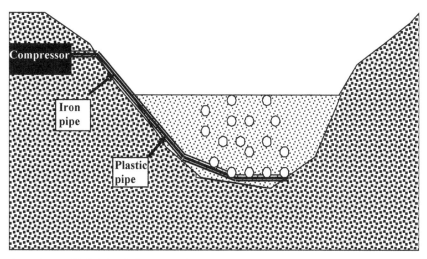

Figure 4.1 Artificial circulation maintains dissolved oxygen levels in threatened estuaries

4.4 DEVELOPMENT AND SPRAWL

Description

Coastal areas have been under enormous developmental pressure in the past several decades. The high value of coastal land is a result of rising demand from a large influx of population into these areas. As the population grows, the amount of impervious surface area and number of buildings increase dramatically. This development often changes the extent of natural vegetation and alters the flow of water through the watershed and can have harmful effects on erosion-sensitive shoreline areas. Development often occurs on waterfront areas such as scenic beaches, often destroying sand dunes and limiting the natural ability of dune movement to buffer storm waves. Activities of large populations also contribute to point and nonpoint source pollution that affects coastal zones. The increased level of pollution caused by development puts a major strain on aquatic habitats, which in turn affects the fishing and tourism in coastal areas.

Problems and Impacts

Nutrient loading Accidental and intentional discharges of wastewater from agricultural land and wastewater treatment facilities cause intense stress and eutrophication in coastal systems. In the early stages of loading, this can be beneficial, yielding larger fish harvests, but beyond a certain threshold, shellfish and other commercial fish can be poisoned. The food chain then becomes disrupted and dissolved oxygen levels become too low to support aerobic organisms.

Increased runoff As development increases the amount of impervious surface area in a watershed, surface waters are unable to infiltrate into ground aquifers. Erosion from runoff increases and the amount of groundwater recharge decreases.

Base flow reduction As demands for drinking water from ground aquifers increase, the water table drops. As a result, the base flow in streams decreases, and the streams become more susceptible to drying out.

Maximum daily yield reached: Interbasin water transfers or desalination options When the daily safe yield from a coastal aquifer is exceeded, the aquifer will draw in marine water which pollutes the aquifer and makes groundwater unusable. The options for obtaining freshwater are to draw in water from another watershed, a very ecologically questionable solution, or else desalinate marine waters, a very expensive alternative.

Public access to the coast A critical issue in coastal areas is the balance between conserving important habitat and allowing the public to utilize the same areas for recreation. In the U.S. large areas of seashore are being purchased by the National Park Service to ensure proper management that is impartial to local economic trends for these areas.

Open space values lost Open, undeveloped space has both aesthetic and ecological value. Open park space is used for a variety of recreational activities as well as serving as land that will allow infiltration of stormwater. Undeveloped land is useful in reducing the volume and speed of surface runoff.

Power plants Production of power that is needed for increasing populations can yield a variety of pollution problems depending on the method of production. Power plants commonly produce air pollution as well as thermal pollution in waterways.

Case Study

Watershed Name: Balikpapan Bay Watershed
Location: Indonesia, Southeast Asia
Major Problem: Coastal watershed protection
Approaches: Integrated plan involving continuous links among stakeholders, awareness programs, and locally tailored partnerships.
Information:
http://www.crc.uri.edu/download/USA_004C.PDF
http://www.usaid.gov/our_work/environment/water/case_studies/indonesia.balik
papan_bay.pdf

4.5 SEDIMENT LOADING

Description

Sediment loading refers to the introduction of significant quantities of sediment into coastal ecosystems. The two primary sources of sediment loading are discharges from rivers into coastal estuaries and deltas, and the sediment deposits from dredging activities. In many coastal areas, sediment loading is a significant problem that impacts human health, recreation, and commercial activities. Sediment loading also causes serious problems in coastal ecosystems and impacts marine life. Certain toxic chemicals and heavy metals in water tend to bind to sediment particles, allowing them to collect in sediments and persist until released when sediments are disturbed (U.S. EPA 1996a).

Problems and Impacts

Outwash Sediments that wash into rivers through runoff and end up as coastal outwash can extend deltas and fill estuaries. The impacts on human health include contamination of shellfish beds by toxic metals, which can cause closure of economically important mariculture operations.

Dredging Sediments from the ocean bottom are often dredged, usually for the purposes of developing shorelines, repairing storm damage, or deepening shipping lanes. Dredging releases sediment into the water column, especially when it is redeposited. The deleterious impacts on water quality and marine life include increased turbidity, the freeing of sediment-stored contaminants, and the burial of marine organisms.

Solutions

Nonstructural

U.S. EPA Contaminated Sediment Strategy The EPA is developing a program to address the issue of sediment contamination (U.S. EPA 1998). The program aims to 1) prevent the volume of contaminated sediment from increasing; 2) reduce the volume of existing contaminated sediment; 3) ensure that sediment dredging and disposal are managed in an environmentally sound manner, consistent with the needs of waterborne commerce; and 4) develop scientifically sound management tools for use in pollution prevention, source control, remediation, and dredged material management.

4.6 NEGATIVE IMPACTS ON MARINE LIFE

Description

The protection of coastal ecosystems is of vital importance to marine life forms. Negative effects on water quality and habitat result in diminished species vigor and reproduction as well as disruptions of intricate food webs. Detrimental biological effects translate into diminished economic and ecological services to human beings in terms of recreation and commerce.

Problems and Impacts

Declining eelgrass beds Eelgrass beds are critical to the early life cycles of many marine organisms - especially fish - because they serve as nurseries where immature organisms can find food and shelter. These aquatic plants are susceptible to disease from pollution, suffocation by eutrophication, and displacement by development and recreational activities. Large-scale die-offs of eelgrass beds in major estuaries along the eastern U.S. seashore have been reported in recent years. The die-offs are significantly disrupting the ecological balance of estuaries.

Shellfish bed contamination Because bivalve shellfish are filter-feeders, they concentrate pollutants, contained in water and sediments, in their stomachs. Although these pollutants do not harm the shellfish themselves, shellfish bed closures are necessary when contaminants are present because of the public health risk. People who consume shellfish raw or improperly prepared are susceptible to gastroenteritis, a type of food poisoning that causes nausea, diarrhoea, abdominal cramps, and Hepatitis-A. According to the U.S. EPA (1996a), 6.7 million acres of shellfish-growing waters in the nation are restricted nationally due to sewage, heavy metal, and toxic chemical contamination. Sewage contamination is the most common reason for shellfish bed closures and is usually detected by testing waters for the presence of E. coli bacteria. E. coli indicates the probable presence of pathogens. Shellfish bed closures can have serious economic repercussions on the seafood and aquaculture industries, especially if public confidence in the quality of the nation's marine food supply is diminished.

Fisheries Inshore fisheries are subject to a variety of problems related to water and habitat quality and human resource use. Commercial over-fishing of species can cause short-term economic losses and threaten the long-term viability of fisheries. When commercial over-fishing occurs, tensions between commercial and recreational anglers can increase. The unintentional harvesting of non-target species can also result in the death and loss of these resources and threaten fisheries viability.

Fish consumption advisories Fish and shellfish accumulate contaminants such as mercury, PCB's and other toxic metals. As these contaminants move up the food chain, they can be unsafe for human consumption. Although there were advisories for 36 different pollutants in 2004, primary bio-accumulative pollutants in the U.S. include mercury, PCBs, chlordane, dioxin, DDT and metabolites (U.S. EPA 2004).

Solutions

Nonstructural

Contaminant testing and fish advisories The U.S. EPA is taking the following steps to address the issue of sediment contamination and fish advisories: 1) completing a national survey intended to identify levels of contaminants found in fish and shellfish; 2) developing nationally consistent processes for monitoring water quality and fish tissue; and 3) reviewing guidelines for decision making on the issuance of fish consumption advisories.

Public education about the health risks associated with exposure to contaminants in locally caught fish and shellfish is an important part of dealing with this issue and minimizing risk. The U.S. EPA has developed a brochure in Spanish and Asian languages for distribution to the public through paediatricians, obstetricians, and health care organizations.

Case Study

Watershed Name: Chesapeake Bay Watershed
Location: New York, Pennsylvania, Maryland, Delaware, Virginia and West Virginia and the District of Columbia, eastern U.S.A.
Major Problem: Watershed and bay ecosystem protection
Approaches: Multi-jurisdictional partnership to protect and restore the bay.
Information: http://www.chesapeakebay.net/index.cfm

4.7 BEACH CLOSURES

Description

The closure of recreational swimming beaches due to water quality issues and public health concerns continues to be a common problem in the United States. According to the U.S. EPA (1998), during 1996 more than 2,500 beaches were either posted with warnings or closed for at least one day because of public health concerns related to water quality.

Problems and Impacts

Bacteriological contamination Sewage pollution originating from sewage overflows, polluted stormwater runoff, boating wastes, and malfunctioning septic systems remains the most common reason for beach closures.

Swimming or ingesting water contaminated with sewage can result in a number of illnesses, including rashes, gastrointestinal upset, hepatitis, and ear, skin, and respiratory infections. Testing for the presence of E. coli is the most effective and commonly used method to determine whether or not beaches are affected with sewage.

Medical wastes Though far less common, the appearance of medical wastes washed up on beaches after being dumped offshore is a serious public health concern. The appearance of syringes, hypodermic needles and other wastes are obvious symptoms of this problem.

Solutions

Nonstructural

In 1998, the U.S. EPA took the following actions to address the issue of beach closures (U.S. EPA 1998):
- Released an internet-based, federal database on beach advisories and closings in the United States.
- Released a BEACH Action Plan that describes priority actions for federal, state, tribal, and local implementation of beach monitoring and notification programs including priority research, training, and guidance needs for the implementing agencies.
- Created a specific plan and schedule for the development of a new generation of microbiological criteria for national beach water quality standards.

4.8 MARINAS AND BOATING

Description

Recreational boating is growing in popularity along coastal areas. Increased levels of recreational boating put habitat and other marine resources at risk (U.S. EPA 1993). The expanding development of marinas along coastal waterways and the resulting pollution from motorized boats can cause water quality and habitat degradation. These impacts have serious ramifications for human health, fragile aquatic ecosystems and marine organisms. The pollutants generated by marinas and boating activities can produce sub-lethal and lethal toxic conditions in the water column. The primary concerns are low dissolved oxygen and elevated levels of metals and petroleum hydrocarbons (U.S. EPA 1993).

Problems and Impacts

Marinas Located at the water's edge, marinas pose unique environmental quality problems, largely because there is no buffering of nonpoint source pollutants before they reach the water body. Among the most important detrimental environmental impacts associated with marinas and boating are: poorly flushed waterways that are susceptible to dissolved oxygen deficiencies; pollutants discharged from boats; runoff of pollutants from parking lots, roofs, and other impervious surfaces; alteration or destruction of wetlands and shellfish areas as a result of the construction of marinas, ramps, and related facilities; and the addition of pollutants generated from boat maintenance activities on land and in the water (U.S. EPA 1993).

Sewage and low dissolved oxygen In addition to being aesthetically unpleasant, untreated sewage that is illegally discharged by recreational boats degrades water quality and introduces harmful bacteria and pathogens into water, particularly fecal coliform. The U.S. Coast Guard has approved the use of three Marine Sanitation Devices (MSDs), or "heads", used to treat sewage. Types I and II MSDs utilize chemical treatment and maceration which allows waste to be discharged into coastal waters, while Type III MSDs are holding tanks which must be discharged more than three miles from shore. Recreational and commercial vessels over 65 feet in length are required by the U.S Federal Water Pollution Control Act of 1972 to use MSD types II or III. Vessels with MSD types I and II can still contaminate waters because the chemicals used to deodorize and disinfect this sewage (alcohol, formaldehyde, zinc, and ammonium salts) degrade water quality (Massachusetts Bays Program 1996.).

Coastal areas are especially prone to water quality problems from human wastes because of high bacteria levels that force the closure of economically important shellfish beds. Sewage discharged from boats requires oxygen to decompose, both while in the water column and when settled into the bottom sediment. This increased demand for oxygen reduces the total level of available oxygen in the water column. In temperate regions this condition poses problems because boating activity peaks during the summer months when water temperatures are highest, oxygen solubility in water is lowest, and the metabolic rates of aquatic organisms are at their highest. With poor flushing conditions in the water body surrounding the marina, which is common, the problem is exacerbated (U.S. EPA 1993).

Metals Because metals and metal-containing compounds are used in boat operation, maintenance, and repair, they are commonly found in the surrounding water column in amounts that are toxic to aquatic organisms. Some of the most prevalent metals used in the boating industry are: lead - used as a fuel additive and released through incomplete combustion and boat bilge; arsenic - used in paint, pesticides, and wood preservatives; zinc - used in anodes to deter corrosion of metal hulls and engine parts; and copper and tin - used as biocides in antifouling paints. These metals settle into bottom sediments and are ingested by many marine organisms, especially shellfish.

Petroleum hydrocarbons The fueling of motorized boats at marinas, the pumping of bilges, and ruptured fuel lines can all lead to the release of petroleum hydrocarbons into waters. Many of these compounds, especially polynuclear aromatic hydrocarbons (PAHs), tend to adhere to particulate matter and become incorporated into bottom sediments, especially under poor flushing conditions (U.S. EPA 1993).

Shoaling and shoreline erosion along coastal areas occur through the physical transport of sediment due to waves and/or currents. Erosion can result from numerous causes, including naturally occurring wave action and currents, and the human-induced effects of boat wakes, which erode waterway banks and beds, or from channel dredging of deposited sediments. Factors influencing the degree of impact from boating traffic include the distance of the boat from shore, boat speed, side slopes, sediment types, and depth of the waterway. The impact on aquatic life is largely from the washing away of aquatic organisms and fringing plants.

Solutions

Structural

Marina management guidelines
- Site and design new marinas such that the bottom of the marina and the entrance channel are not deeper than adjacent navigable water unless it can be demonstrated that the bottom will support a natural population of benthic organisms.
- Design new marinas with as few segments as possible to promote circulation within the basin.
- Consider design alternatives in poorly flushed water bodies (open marina over semi-enclosed design; wave attenuations over a fixed structure) to enhance flushing.
- Design and locate entrance channels to promote flushing.
- Establish two openings, where appropriate, at opposite ends of the marina to promote flow-through currents.

- Designate areas that are or are not suitable for marina development; i.e. identify in advance the water bodies that do not have flushing adequate for marina development.
- Establish entry and exit points for streams, lakes, and rivers and implement erosion control practices, e.g. create stone pathways.

Nonstructural

- Boaters should be sure to remove all plants attached to motor props and boat hulls before launching their vessels, and to pump bilge waters before leaving waterways. Public education, e.g. signs located at launch points, can alert boaters to the problem of introduced and invasive species and gain cooperation in preventing the problem.
- The U.S. Federal Clean Vessel Act of 1992 gave assistance for states to set up grant programs for the construction, renovation, operation, and maintenance of boat pump-out and waste reception facilities at public and private marinas (Massachusetts Bays Program 1996).

4.9 OIL POLLUTION

Description

Oil spills of 34 tons or more have occurred in the waters of 112 nations since 1960, with a higher frequency of occurrence in certain hotspots (Etkin 1997). The top five hotspots reported are the Gulf of Mexico (267 spills), northeastern U.S. (140 spills), Mediterranean Sea (127 spills), Persian Gulf (108 spills), and the North Sea (75 spills) (Etkin 1997). While oil spills from commercial vessels are the most catastrophic to the marine environment, industrial and municipal wastewater, stormwater runoff, boats, and creosote-treated wood pilings actually contribute the majority of the pollutant load. Oil pollution adversely impacts much of the marine environment. Oil pollution is especially threatening to stationary plants, sensitive species, and organisms in early life stages, which suffer high mortality immediately after a spill. Those organisms that survive suffer short-term stress and impaired metabolism. Economically and environmentally important marine habitats and communities especially impacted by oil pollution include shellfish and eelgrass beds.

Structural

Dispersants and bioremediation agents are commonly used in oil spill response. While dispersants break up surface oil slicks, bioremediation agents are applied to residual oil on shorelines. Bioremediation acts by enhancing microbial degradation of the compound.

SUGGESTED READING

4.1 – 4.9 Coastal Watersheds

Etkin, D.S. (1997) *Oil Spills from Vessels (1960-1995): An International Historical Perspective,* Cutter Information Corporation, Cambridge, Massachusetts.

Massachusetts Bays Program (1998) *Massachusetts Bays Watershed Stewardship Guide: An Education Resource.* URL: http://www.msp.umb.edu/mbea/mbeaguid.htm.

Massachusetts Bays Program (2003) *Massachusetts Bays Comprehensive Conservation & Management Program: An Evolving Plan For Action.* URL: http://www.mass.gov/envir/massbays/pdf/revisedccmp.pdf.

U.S. EPA (1993) *Guidance Specifying Management Measures For Sources of Nonpoint Pollution In Coastal Waters.* Report EPA 840-B-92-002, U.S. Environmental Protection Agency, Office of Water, Washington D.C.

U.S. EPA (1996a) *Environmental Indicators of Water Quality in the United States.* Report EPA 841-R-96-002, U.S. Environmental Protection Agency, Washington, D.C.

U.S. EPA (1998) *Clean Water Action Plan: Restoring and Protecting America's Waters.* Report EPA 840-R98-001, U.S. Environmental Protection Agency, Office of Water, Washington, D.C.

U.S. EPA (2004) *2004-National Listing of Fish Advisories.* Fish Advisory Program, U.S. Environmental Protection Agency, Washington, D.C. URL: http://www.epa.gov/waterscience/fish/advisories/

U.S. EPA (2006a accessed) *Coastal Watershed Fact Sheets.* URL: http://www.epa.gov/owow/oceans/factsheets/fact5.html.

U.S. EPA (2006b accessed) *Estuaries and Near Coastal Areas, Bioassessment and Biocriteria Guidance.* URL: http://www.epa.gov/ost/biocriteria/States/estuaries/estuaries1.html.

U.S. Fish and Wildlife Service (1970) *National Estuary Study.* U.S. Department of the Interior, Washington, D.C.

5

Biodiversity and Ecosystem Health

Human alteration of a watershed can result in land transformation at many different spatial and temporal scales and in many different kinds of habitats. Detrimental changes to important habitat can negatively impact wildlife populations and ecosystem biodiversity.

Landscapes are defined at different temporal and spatial scales (e.g. regional, watershed), and are often described in terms of matrix, patch, corridor, and mosaic components (Figure 5.1).

- *Matrix* is the land cover that is dominant and interconnected over the majority of the land surface. Often the matrix is forest or agriculture, but theoretically it can be any land cover type.
- *A patch* is a nonlinear area that is less abundant than, and different from, the matrix.
- *A corridor* is a special type of patch created by the connection of patches in the matrix. Typically, a corridor is linear or elongated in shape, such as a stream corridor.
- *A mosaic* is a collection of patches, none of which is dominant enough to be interconnected throughout the landscape (FISRWG 1998).

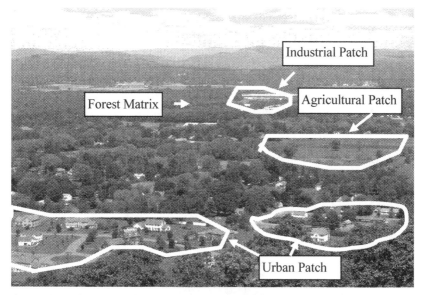

Figure 5.1 Landscape mosaic: patch, corridor and matrix

5.1 LANDSCAPE EFFECTS

Six major types of land transformation are common and widespread: deforestation, suburbanization, corridor construction (human-made), desertification, agricultural intensification, and reforestation (Forman 1995). Landscape transformation can result in the extinction of native species, loss of biodiversity, the disruption of ecological processes (e.g. food webs), and facilitate the introduction of invasive species. These conditions often result in serious environmental and ecological impacts upon both terrestrial and aquatic systems. Maintaining the ecological health of a watershed landscape requires managing entire watersheds, including private and multiple-use lands, so as to minimize destruction and isolation of natural habitats.

5.1.1 Landscape Transformation

Description

Natural disturbances and processes create heterogeneous landscapes rich in native species, while human land uses often create islands of natural habitat embedded in hostile matrices. The landscape initially starts with natural vegetation. Disruption of extensive habitats through human activity most commonly begins with the formation of gaps or perforations in the vegetative matrix. Examples are utility line trails, access roads through woodlands, or the perforation of forests by residential development. The initial impact upon species composition and abundance is generally minimal, but as gaps get bigger and increase in number, the matrix shifts to a fully fragmented landscape (Figure 5.2).

Habitat changes may reduce the number and abundance of native species. In addition, transformed landscapes reduce or prevent normal wildlife dispersal patterns. Dispersal plays a critical role in long-term population viability of species, wildlife migration, species extinction, and the introduction of nonnative species. Landscape transformation reduces biodiversity, the variety of habitats, and leads to ecosystem decline. The following species-scale effects are generally seen to increase with fragmentation: isolation; number of generalist, multi-habitat, and edge species; number of exotic species; nest predation; and extinction rates (Forman 1995).

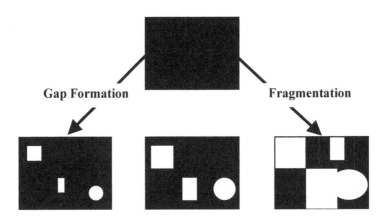

Figure 5.2 Landscape gap formation and fragmentation

Predicting the effects of fragmentation on wildlife and habitat is difficult because the outcome is influenced by multiple factors, including the life history of the native species or of potential introduced species and the ecology of the habitat. Characteristics that make species especially vulnerable to the effects of landscape changes include: rarity, large home ranges, limited power of dispersal, low reproductive potential, dependence on resources that are unpredictable in time or space; ground-nesting, occupancy of habitat interiors and persecution or exploitation by people (Meffe and Carroll 1994).

Problems and Impacts

Landscape Scale

Fragmentation is the breaking up of a habitat into smaller parcels. When used in the general sense, fragmentation can be viewed as including the following spatial processes:

Perforation is the making of holes in the landscape. The most common origin of land transformation can be either natural (e.g. tree blow down) or caused by humans (e.g. logging).

Dissection is the subdivision of an area using equal-width lines, such as a road network.

Shrinkage is a decrease in the size of the object, such as a patch. This is a universal characteristic of land transformation.

Attrition is the disappearance of objects, such as patches (Forman 1995).

Species scale

Initial exclusion is the elimination of species that existed only in portions of the impacted habitats. Many rare species are endemic with very narrow geographic distributions.

Barriers and isolation When habitats are isolated because of the establishment of barriers to movement, species are adversely impacted. In cases where one single patch is not enough to support individuals or a species group, the patch is vulnerable to population decline and eventual extinction.

Island-area effect Small fragments are more susceptible to extinction because they contain fewer habitats and support smaller populations of native species.

The *crowding effect* occurs when an area is isolated by the destruction of surrounding natural habitat. The initial population densities of animals increase as a result of displacement, followed by a collapse due to overpopulation.

Biodiversity loss can change gene frequencies within a population, resulting in the genetic weakening of adaptive traits. When a small population is isolated from the rest of the population, its susceptibility to disease and environmental changes increases. Changes in the distributions of species and ecosystems can occur locally, at the ecosystem level, or can extend to a continental level.

Food chain disruption The displacement or extinction of species from habitats and ecosystems as a result of fragmentation can have devastating effects upon the food webs of those communities.

Exotic species are introduced invaders that colonize disturbed habitats. Biological invasion depends on the particular characteristics of individual species, the potential host community, and the presence or absence of abiotic factors. Native species are often out-competed and eliminated as a result of exotic species introductions. All species introductions and losses are not equal. A decline in biodiversity can result from the elimination of sensitive species even though the overall richness or number of species may remain the same or even increase (Meffe and Carroll 1994).

Solutions

Structural

Retain connectivity Identifying and protecting critical habitats and maintaining connectivity of landscapes lessen the impact of fragmentation in watersheds. Retaining landscape connectivity facilitates wildlife movement, reproduction, and preserves habitat suitability. It is a critical step in assuring the existence and abundance of species. Possible management practices include:
- Establishment of buffer zones with low intensity land use around natural areas to minimize edge effects.
- Maintenance of natural vegetation along riparian areas in strips as wide as possible to minimize edge effects.
- Minimization of disturbed habitats in order to reduce the potential for biological invasions.

Nonstructural

Landscape analysis and planning in order to:
- Identify the pattern of habitats and connections as they relate to the needs of native species. Determine where the major, unfragmented blocks of habitat exist.
- Identify corridors and natural connections between habitats that can be maintained.
- Determine the significance of the landscape, e.g. conservation value, in the larger context of regional, national, and global scales.
- Avoid further fragmentation or isolation of habitats through planning and management, e.g. clustering development to maintain connectivity.

- Identify and protect traditional wildlife migration routes, thereby minimizing human influence and disturbance.

5.1.2 Natural Disturbances

Description

Natural forces are constantly affecting and shaping ecosystems. Sometimes these forces are expressed as major disturbances, events that significantly alter the pattern of variation in the structure or function of a system (Forman, 1995). Major disturbances are normal in an ecological system, although they are not generally frequent. Sometimes these disturbances spread within and between ecosystems. Disturbances influence probabilities of extinctions and colonization and thus affect the pattern of biodiversity in the landscape (Meffe and Carroll 1994). An important concept is the intermediate disturbance hypothesis, which states that maximum species richness in many systems will occur at an intermediate intensity and frequency of natural disturbance (Meffe and Carroll 1994).

This hypothesis posits that: 1) high disturbance levels allow persistence of only those species that are disturbance-adapted; 2) low disturbance levels allow competitive dominance by some species, causing local extinctions of others; and 3) intermediate disturbance levels allow the coexistence and persistence of many species, especially in patches of different disturbance types and intensities.

Problems and Impacts

Fire Wildfires can be an essential natural process when they occur at normal frequencies and intensities (Meffe and Carroll 1994). Wildfires are a part of the rejuvenation process in many types of ecosystems. For example, some tree species are dependent upon fire for reproduction while others benefit from the clearing process. Additionally, while wildlife may be dislocated by the fire event, many species are drawn to wildfire sites once new vegetation starts to appear. However, increased development in naturally fire prone areas leads to increased pressure to intervene in this process.

Water Floods resulting from heavy rains and hurricanes can result in a myriad of ecosystem effects. These effects range from minor stream alterations and the creation of water quality problems, to tremendous loss to society including property destruction, and incidence of injuries and deaths. Floods may permanently alter natural systems, but they generally adapt to the changes over time.

Windstorm Damage to natural systems from windstorms includes massive tree blowdowns in forests, large-scale displacement of sand and agricultural soils, and significant loss of life and property.

Volcanoes Full-scale volcanic eruptions are capable of obliterating landscapes and destroying ecosystems. The impacts of massive explosions (such as Mt. St. Helens in the U.S.) include lava flows and airborne dust that can obliterate and permanently alter landscapes. Plant species may take several years to reestablish themselves; wildlife species may take longer to return to affected areas.

Case Study

Watershed Name: Gokulpura and Govardhanpura watershed
Location: Bundi district, Rajasthan, India
Major Problem: Drought
Approaches: Watershed management through conservation practices that include field bunding, afforestation, gully plugs, checkdams, underground bandharas, and gabion structures.
Information:
http://www.fao.org/ag/aGL/watershed/watershed/papers/papercas/paperen/case14en.pdf

---

5.2 LANDSCAPE PLANNING

Landscape planning requires the comprehensive assessment of a watershed that includes spatial and temporal contexts, knowledge, flexibility, and collaboration. Information on watershed structure and function, the structure and pattern of the landscape, flows and movements of matter and energy, and rates and directions of change in the watershed system should be part of an effective planning process.

5.2.1 Planning Landscape Components

Description

Planning components that need to be thoroughly evaluated in watershed management are discussed below:

Context Different landscapes serve as sources and sinks for water, so their spatial arrangement is an important aspect of planning. The temporal context should also be considered; for example, the regimes of natural patterns and human plans at 5, 10, and 20 year intervals (Forman 1995).

Landscapes require maintenance of four priority components (Forman 1995):

- A few large patches of natural vegetation;
- Wide corridors along streams;
- Connectivity for movement of key species among the large patches; and
- Maintaining heterogeneous bits of nature throughout human-developed areas.

Key locations that can be targeted for protection (Forman 1995)

- Unusual features that can act as sources and/or sinks of energy and material.
- Elements with the highest species richness.
- Large nodes in networks for their connectivity in landscape.
- Fix gaps in major corridors that are detrimental to movement along, and flow through, the corridor.
- Elements sensitive to human impact.
- Focus on strategic points, permitting important control over flows and protection from external forces.

Ecological characteristics that can be targeted (Forman 1995)

- Maintain healthy populations of migratory fish.
- Maintain viable populations of important species.
- Minimize inbreeding depression.
- Plan for the needs of multi-habitat species.
- Plan for wide ranging species.
- Avoid the spread of exotic species.
- Protect keystone species whose decimation affects so many others.
- Design to concurrently enhance interior and edge game species.
- Channel snow accumulations to desirable locations.
- Protect best soils from overuse or covering by buildings.

Spatial attributes (Forman 1995) that can be targeted in planning are: 1) patch and boundary attributes (small patches; minimum dynamic area; soft, curvilinear mosaic boundaries; edges; and compact shapes); 2) corridor attributes (gap size and number; clusters of stepping stones; major road corridors as conduits, barriers, and sources; and locations such as gaps and the heads of valleys); 3) network, matrix, and mosaic attributes (connectivity and circuitry of a network or matrix; and perforation of the matrix); and 4) flows and movements over the landscape (sources and sinks for heat, gases, and materials) (Forman 1995).

5.2.2 Stream Corridor Management

Description

A stream corridor is an ecosystem that usually consists of a stream channel, floodplain, and transitional upland fringe (Figure 5.3). Together these different components function as a dynamic and valuable crossroad in the watershed landscape (FISRWG 1998). Stream corridors provide many important hydrologic, hydraulic, geomorphic, and other ecological functions:

"Water and other materials, energy, and organisms meet and interact within the stream corridor over space and time. This movement provides critical functions essential for maintaining life such as cycling nutrients, filtering contaminants from runoff, absorbing and gradually releasing floodwaters, maintaining fish and wildlife habitats, recharging groundwater, and maintaining streamflows." (FISRWG 1998).

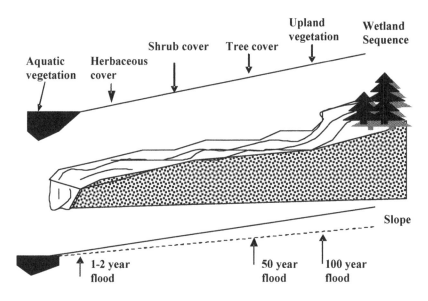

Figure 5.3 Stream and river corridors

Along with providing home range, stream corridors act as conduits for migration, dispersal, and relocation of wildlife. Since species diversity and richness is dependent upon the healthy ecological functioning of a stream corridor, protection of all of the components of the stream corridor is important.

Determining the optimal width of a stream corridor involves three steps: 1) delineation of the key ecological processes or functions performed by the corridor; 2) separating the basic types of stream and river from headwaters to mouth, based on the spatial structure within the corridor; and 3) determining the corridor width needed by each stream or river type, by linking the most sensitive ecological processes with the spatial structure (Forman 1995). Land value plays an important role in policies related to stream corridor protection.

Management Approaches

Structural

Instream practices that could be adopted include (FISRWG 1998):

Boulder clusters Groups of boulders that are placed in the base flow channel to provide cover, scour holes, or areas of reduced velocity.

Weirs or sills Log, boulder, or quarry stone structures that are placed across the channel and anchored to the stream bank to create pool habitat, control bed erosion, or to collect and retain gravel.

Fish passages Instream changes that enhance the opportunity for a target fish species to freely move upstream to spawning areas, utilize habitat, and perform other life functions.

Log/brush/rock shelters Logs, brush, and rock structures can be installed in the lower portion of stream banks to enhance fish habitat, encourage necessary food web dynamics, prevent stream bank erosion, and provide shade.

Lunker structures Cells constructed of heavy wooden planks and blocks that are embedded into the toe of stream banks at a channel level, which provide covered compartments for fish shelter, habitat, and prevention of stream bank erosion.

Migration barriers Obstacles placed at strategic locations along streams to prevent undesirable species from accessing upstream areas.

Tree cover Felled trees placed along the stream bank can provide overhead cover, aquatic organism substrate and habitat, stream current deflection, and scouring, deposition, and drift catchments.

Wing deflectors Structures that protrude from either stream bank but do not extend entirely across a channel. They deflect flows away from the bank, and scour pools by constricting the channel and accelerating flow.

Stream bank practices that could be used (FISRWG 1998):

Live cribwalls Fencing which minimizes erosion losses.

Brush mattress A combination of live stakes, live fences, and branch cuttings can be installed to cover and physically protect stream banks. The brush mattress will eventually sprout and establish numerous individual plants.

Joint planting Live stakes tamped into the joints or openings between rocks can be installed on a slope face.

Bank shaping and placement This includes regrading stream banks to a stable slope, placing topsoil and other materials needed for sustaining plant growth, and selecting, installing, and establishing appropriate plant species.

Branch packing Alternate layers of live branches and compacted backfill can be used to stabilize stumps and revegetate holes in stream banks.

Stone toe protection A ridge of quarried rock or stone cobble can be placed at the toe of the stream bank as armour to deflect flow from the bank, stabilize the slope and promote sediment deposition.

Tree revetments A row of interconnected trees attached to the toe of the stream bank or to deadmen in the stream bank can be used to reduce flow velocities along eroding stream banks, trap sediment, and provide a substrate for plant establishment and erosion control.

Riprap is a blanket of appropriately sized stones extending from the toe of a slope to a height needed for long term durability.

Livestock exclusion Use fencing, provide alternate water and shelter, and manage grazing to protect, maintain, or improve riparian flora and fauna and water quality.

Forest riparian buffers provide streamside vegetation to lower water temperatures, supply a source of detritus and large woody debris, improve habitat, and reduce sediment, organic material, nutrients, pesticides, and other pollutants migrating to the stream.

Other measures that can be used include log, rootwad, and boulder revetments, dormant post plantings, and vegetated gabions.

Case Study

Watershed Name: Motueka River Watershed
Location: Northern end of South Island, New Zealand
Major Problem: Conflicts over resource management
Approaches: Integrated Catchment Management (ICM) is used to integrate technical knowledge and social learning.
Information:
http://www.fao.org/ag/aGL/watershed/watershed/papers/papercas/paperen/case8en.pdf

---

5.2.3 Open Space Protection

Description

Strategies to protect open space should be an essential component of watershed planning. Open space areas provide important ecological and recreational functions, and are vital to community character and aesthetics. However, development pressure puts these areas at risk, unless communities are able to find creative ways to protect them. Components to be considered in an open space plan include areas of natural vegetation and agriculture, the built environment, and corridors for wildlife, water protection, and human use. Planning for open space protection must include the identification and prioritization of broad patterns, major corridors and special ecological sites in the watershed.

Solutions

Structural

Greenways are linear conservation corridors mainly located near suburban areas. Primarily constructed for recreational or aesthetic purposes, greenways are increasingly being used to preserve ecological functions and wildlife benefits. Funding for greenways can originate from communities or at the state and federal level. Planning and implementation of greenways should involve diverse private and public stakeholders, and community volunteers.

Nonstructural

Cluster developments concentrate buildings to save open space in a location. Many communities are considering cluster development as a way to protect open space. Local zoning codes can be modified to enable more compact development. This special permitting tool allows a builder to construct houses on smaller lots in groups or clusters and requires that the land saved is permanently protected as open space. The overall density of the cluster development is not greater than a conventional development on the same parcel.

Zoning is land use regulation that is a part of the local bylaws. Cluster development is one type of zoning that can be used to protect open space. Zoning places restrictions on land uses, residential lot size, and commercial and infrastructure development.

Easements are legal arrangements whereby landowners receive compensation for allowing public access to their private lands. In the northeastern U.S. this arrangement is made with the town's Conservation Commission.

Acquisition is the process of purchasing rights to the land for the purpose of preservation. This is a common policy when funding is available to compensate the owner for losses incurred.

SUGGESTED READING

5.1 – 5.2 Landscape Effect and Planning
FISRWG (1998) *Stream Corridor Restoration: Principles, Processes and Practices.* Report GPO Item No. 0120-A; SuDocs No. A 57.6/2:EN3/PT.653. ISBN-0-934213-59-3, Federal Interagency Stream Restoration Working Group, Washington, D.C. URL: http://www.nrcs.usda.gov/technical/stream_restoration.
Forman, Richard T.T. (1995) *Land Mosaics: The ecology of landscapes and regions*, Cambridge University Press, New York.
LIH Landscape Information Hub (2006) *Landscape Planning Web: Links.* University of Greenwich, London. URL: http://www.landscapeplanning.gre.ac.uk/links.htm.
Meffe, Gary K. and Carroll, C. Ronald (1994) *Principles of Conservation Biology*, Sinauer Associates, Inc., Sunderland, Massachusetts.

6

Water Assessment:
Quality and Quantity

6.1 WATER QUALITY MONITORING

With growing concern on the effects that microorganisms and pollution have on public health, the need to create and maintain strict standards for water quality has become clear. Monitoring and assessment of water quality is an important part of water quality management. The United Nations GEMS/Water Programme collects water-related data and information on the state and trends of global inland water quality. This information can be used for sustainable management of the world's freshwater supply and to support global environmental assessments and decision-making processes (UNEP 2005b). The U.S. EPA classifies waterways as: *Class A*: drinkable/swimmable, *Class B*: fishable, or *Class C*: boating only. Public awareness of the status of water quality in the watershed can help avoid health and safety problems. Classifications are also used to help with prioritization of restoration efforts.

6.1.1 Temperature

Description

Water temperature is an important water quality parameter that influences the rates of several biological and chemical processes. Increased water temperatures

© 2007 IWA Publishing. *Watershed Management: Issues and Approaches* by Timothy O. Randhir. ISBN: 9781843391098. Published by IWA Publishing, London, UK.

are known to increase biological activity and influence the rate of many chemical reactions. Aquatic organisms are dependent on a range of temperature for optimum health. If temperatures remain outside of this range for a prolonged period of time, organisms become stressed and can die. Temperature also affects the oxygen content of water (lower oxygen content with increasing temperature), the metabolic rate of aquatic organisms, the rate of photosynthesis in plants, and the sensitivity of organisms to toxic wastes, parasites, and diseases. Temperature change is affected by the weather, the removal of shading riparian vegetation, impoundment (e.g. dams), discharge of cooling waters from power plants, urban stormwater, and groundwater inflows (U.S. EPA 1996a).

Water temperature should be monitored when: 1) the potential exists for large changes in water temperature due to management actions and land use activities; 2) water temperatures are already at the upper limit of the acceptable range; and 3) there is a potential for significant temperature increases due to the additive effects of numerous smaller increases (MacDonald et al. 1991).

6.1.2 pH

Description

The alkalinity or acidity of a substance is determined by its pH. The pH is defined as the concentration of hydrogen ions, H^+ and OH-, in water as moles per litre (MacDonald et al. 1991). pH is measured on a logarithmic scale over 14 orders of magnitude. When both types of ions are in equal concentration, the pH is 7.0 or neutral; when pH is below 7.0, the water is acidic (more H^+ than OH^-), and when the pH is above 7.0, water is alkaline (more OH^- than H^+).

The pH of waters is controlled by factors such as the contribution of organic debris from forestry activities, atmospheric deposition, acid rain, surrounding rock weathering, and certain wastewater discharges. The pH affects many chemical and biological processes in water, and thus has direct and indirect effects on water chemistry and the biota of aquatic ecosystems. A large variety of aquatic organisms prefer pH ranges between 6.5 and 8.0 (U.S. EPA 1997b). When pH occurs outside the range of tolerance, it stresses the physiological systems of organisms and can reduce reproduction, e.g. egg production and hatching success in salmonids. The solubility of many metal compounds also varies with pH. A low pH allows toxic elements to become mobile and available for uptake by aquatic organisms. This factor is of critical importance in areas with high levels of heavy metals in bottom sediments and for certain sensitive species, such as the rainbow trout. Areas that often exhibit acidic conditions and which are candidates for intensive monitoring programs include, areas affected by wet and dry acid deposition, high altitude areas which typically have thin soils and little buffering capacity, and mining operation sites (MacDonald et al. 1991).

6.1.3 Dissolved Oxygen and Biological Oxygen Demand

Description

Dissolved oxygen (DO) is an important component of natural water systems. Respiration by aquatic animals, decomposition, and various chemical reactions require oxygen and are thus affected by DO in water. The actual concentration of DO in water is dependent on the presence of oxygen sources (major sources include photosynthesis and the dissolution of atmospheric oxygen in water as oxygen levels are depleted), altitude, and the presence or absence of oxygen sinks (MacDonald et al. 1991). DO levels fluctuate with temperature, both seasonally and on a daily basis. The capacity of water to hold oxygen in solution (DO saturation) is inversely proportional to the water temperature. High water temperatures lead to low concentrations of DO in water at saturation, and increase the rate of biological oxygen demand (BOD).

Biological oxygen demand is oxygen needed by biodegrading materials. Accordingly, stream waters may have low DO saturation in the following situations:

- Very slow, low-gradient streams where the rate of aeration is low.
- Sites where fresh organic debris causes a large BOD.
- Warm, eutrophic streams where high rates of photosynthesis and respiration cause diurnal fluctuations in DO.
- Ponded sites such as those caused by beavers (MacDonald et al. 1991).

Given these factors, aquatic organisms are most vulnerable to lowered DO early in the morning, when aquatic plants have not been producing oxygen since sunset of the previous day, during hot summer weather in streams with low flows and high temperatures (U.S. EPA 1996a).

Thus, monitoring for DO should be considered under: 1) low water flow with high temperature; 2) where the rate of energy dissipation, which accelerates re-aeration, is low; and 3) where oxygen loss is high.

6.1.4 Nutrients

Phosphorus is found in two forms in water, dissolved and particulate. In the dissolved form it is found as phosphate ions, which readily bind with other chemicals. Three main classes of phosphate compounds - orthophosphates, condensed phosphates, and organically bound phosphates - are soluble and available for binding with other particulate matter, while only orthophosphates are available for uptake by biota (MacDonald et al. 1991) (Figure 6.1).

Phosphorus can be a limiting factor and is a major player in the eutrophication process (see also Chapter 3).

Nitrogen is found in several forms (Figure 6.1). Nitrate (NO_3) is a highly soluble, stable compound that can dissolve easily, and be readily transported in streams and groundwater. Nitrite (NO_2) is relatively short-lived in water, and is quickly converted to nitrate by bacteria. Ammonia is an inorganic form of nitrogen that is the least stable in water. Conversion of ammonia to nitrite by *Nitrosomonas* and then to nitrate (through *Nitrobacter*) takes place through the process of nitrification.

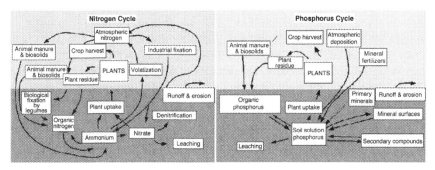

Figure 6.1 Nutrient cycles

6.1.5 Pathogens

Description

The bacterial pathogens affecting water quality include total coliform (TC), fecal coliform (FC), fecal streptococci (FS), and Enterococci. High counts of TC, FC, or FS are usually associated with inadequate sewage treatment, poorly functioning septic tanks, high concentrations of animals, improperly managed recreational areas, and inadequate disposal of wastes.

Total coliform includes a variety of aerobic, facultative, and anaerobic bacteria. Many are nonpathogenic and are not associated with human waste. Total coliform (TC) has been traditionally used as the primary criterion for the sanitary condition of domestic water supplies because of the positive correlation between TC and the incidence of gastrointestinal disease. Coliform bacteria are found in the intestines of humans and animals and in feces, so TC tests may indicate water contamination. Since any bacterial contamination of drinking water is considered unacceptable, TC is used as an indicator for assessing contamination of drinking water because it is the largest and most diverse bacterial group.

Fecal coliform is mostly present in the gut and feces of warm-blooded animals. Fecal coliform (FC) are less able to survive in natural waters than non-fecal coliform bacteria.

Fecal streptococci are found in the intestines of humans and animals. Fecal streptococci are less common than fecal coliform in animals, because they have a more limited ability to survive in natural waters than FC.

Enterococci, is a part of the larger FC group and is a better indicator of the risk of gastrointestinal illness than TC, FC, or FS. The use of this test is becoming more common because it is more sensitive and provides a better estimate of the human health risk. FC, FS, and Enterococci are more specialized groups of bacteria than TC, and their concentration can be more directly linked to sanitary water quality and human health risk.

6.1.6 Turbidity

Description

Turbidity is a measure of the cloudiness of water and refers to the amount of light that is scattered or absorbed by a fluid (MacDonald et al. 1991). Turbidity serves as a measure of water clarity, with increasing turbidity described as a visual increase in cloudiness. Common causes of turbidity include the presence of suspended particles of soil and clay, algae, plankton, fine organic matter, colored organic compounds, and microbes. Erosion, waste discharge, urban runoff, eroding stream banks, and excessive algal growth are all sources of turbidity. Turbidity is an important water quality parameter for aesthetic reasons, and it has a direct detrimental effect on the recreational appreciation of water.

The biological impacts of turbidity include:
- Increased water temperature due to heat absorption by suspended particles.
- Reduced penetration of light, resulting in reduced photosynthesis and decreased primary productivity (especially periphyton and attached algae).
- Higher turbidity can reduce light penetrating the water, and it reduces production of DO.
- Negative impacts on fish; clogged gills and damage to gill tissue, reduced resistance to disease, lowered growth rates, and smothering of fish eggs when sediments settle.

Turbidity is a quick indicator of the effects of runoff from land use activities such as logging, agriculture, and construction. It is also a good indicator of stream erosion caused by stormwater flows in areas of high imperviousness. Although turbidity measurements provide an indication of the amount of suspended material in the water, the precise relationship between turbidity and the mass of suspended material is dependent upon the size and type of suspended particles and includes the measurement of total solids (MacDonald et al. 1991).

Turbidity is generally measured through the use of a turbidity meter. A Secchi disk or transparency tube can be used to measure the transparency, not turbidity, of a lake or slow moving river.

Case Study

Watershed Name: North Santiam Watershed
Location: Oregon, U.S.A.
Issue: Water quality (persistent turbidity)
Approaches: Federal and state partnership to coordinate protection efforts.
Information:
http://or.water.usgs.gov/proj/or00311/proj_desc.html
http://www.fsl.orst.edu/wpg/pubs/Muddy%20Waters%20full.pdf
http://www.epa.gov/safewater

6.2 BIOLOGICAL MONITORING METHODS

Description

Chemical testing is a common method to assess water quality and continues to be accurate in assessing water degradation. However, the long term protection of water resources also requires the monitoring and protection of the living biota of aquatic ecosystems for the following reasons: 1) chemical testing by itself fails to explain the widespread losses and other detrimental trends currently being experienced by aquatic organisms; and 2) humans rely on living waterways for essential goods and services other than water quality and quantity. Waters that cannot support healthy biological communities will most likely not be able to support human society (Karr 1997).

"Biological monitoring and assessment ultimately provide objective descriptions of the conditions of our waters. They diagnose and integrate chemical, physical, and biological impacts as well as their cumulative effects;

they can serve many kinds of environmental and regulatory programs when integrated with chemical analyses and single-species toxicity testing in the laboratory; and they are cost effective. Most importantly, because biological monitoring is sensitive to different impacts of human activity, it offers greater protection to water resources than chemical testing alone can provide" (Karr 1997).

6.2.1 Species Indicators

Indicator species are those organisms that are used to assess the condition of a particular habitat, community, or ecosystem. As it may not be possible to include all species in management plans, biodiversity planning can be achieved by focusing on the management of particular species that are sensitive to habitat fragmentation, pollution, and/or other environmental stressors that affect biodiversity. These indicator species, therefore, serve as surrogates for the larger community (Meffe and Carroll 1994). The species indicator concept has dominated biological evaluations and has relied on ecological studies that typically focused on a limited number of parameters (e.g. species distribution, abundance trends, standing crop, and production estimates). These parameters may or may not be reflective of overall ecosystem health.

6.2.1.1 Aquatic Species Indicators

Periphyton Aquatic flora can be classified into one of three primary categories: 1) free-floating or planktonic forms; 2) plants attached to the substrate; and 3) plants rooted in the substrate. Studies have indicated that the attached plant community is best suited to water quality monitoring (MacDonald et al. 1991). Periphyton is a complex matrix of algae, cyanobacteria, heterotrophic microbes, and detritus.

Diatoms, the most important and diverse algal group in benthic communities, are commonly used in monitoring for water impairment. Periphyton is the base of most aquatic food webs and the source of primary productivity. Mats of benthic algae form rich assemblages of plant, bacteria and animal species. Increased benthic algal production is linked to the increased production of benthic invertebrates and fish (MacDonald et al. 1991).

Advantages to using algae as indicators include:

(1) The presence and growth of algae integrate numerous physical factors.
(2) The relatively short life cycle of algae make them useful indicators of short-term impacts.
(3) Algae are sensitive to certain pollutants, such as herbicides and excessive inputs of nutrients, which may not affect other organisms.
(4) Sampling can be easy and inexpensive.

(5) Relatively standard methods exist for evaluating the structural and functional characteristics of algal communities.

Disadvantages to the use of algae (and aquatic plants) as indicators include:
(1) Algae are highly variable with location.
(2) Algae are highly sensitive to small changes in current, velocity, and other physical factors.
(3) Considerable expertise and time are needed to identify both attached and free-floating microflora species.
(4) The use of qualitative information, such as the presence of particular species, may be invalid or appropriate only on a very coarse scale.

Macroinvertebrates are animals that do not have backbones and are large enough to be seen with the naked eye. They include annelids, crustaceans, flatworms, molluscs, and insects. Benthic, or bottom-dwelling macroinvertebrates, are most often used as a tool for rapid assessment of water quality problems and to classify aquatic habitats according to a variety of water resource criteria. Benthic macroinvertebrates (BMI) posses the following characteristics which make them useful as indicators of water quality: 1) limited migration patterns or a sessile mode of life that is conducive to the study of site-specific impacts; 2) life spans of several months to a few years allow them to serve as indicators of past environmental conditions; 3) abundance in most streams; 4) relatively easy sampling in terms of time and equipment; and 5) sensitivity to habitat and water quality changes make them more effective indicators of stream impairment than chemical measurements (MacDonald et al. 1991). An evaluation of the presence or absence of benthic macroinvertebrates is a useful rapid assessment tool, requiring little effort in comparison to quantitative approaches. In addition, the results of site surveys can be expressed as a single score based on several impact related measures, and/or by using environmental quality categories (DeShon 1995).

Fish communities are highly visible and sensitive to freshwater ecosystem changes. Fish respond predictably to changes in abiotic factors such as habitat and water quality, and have several attributes that make them useful indicators of biological integrity and ecosystem health (Simon and Lyons 1995).
    Advantages to using fish as indicators include:
(1) The mobility and relatively long life span of fish allow them to indicate broad-scale and long-term habitat conditions.
(2) The higher trophic position occupied by fish means that they can be used as an indicator of changes in the lower trophic levels.
(3) Fish are relatively easy to collect and identify in the field.

(4) The habitat requirements of many species of fish are relatively well known.

Disadvantages of using fish as indicators include the following:
(1) The difficulty of obtaining a representative sample or an accurate estimate of the population.
(2) The variety of extraneous factors that can affect fish populations during different life history stages.

Examples of quantitative indices based on fish communities include indicator species or guilds, species richness, diversity, similarity indices, the Index of Well-Being, multivariate ordination and classification, and the Index of Biotic Integrity (IBI). The IBI method is the most commonly used and widely applied. The original version of the fish IBI utilized 12 metrics that reflected species richness and composition, number and abundance of indicator species, trophic organization and function, reproductive behaviour, fish abundance, and condition of individual fish. The original version has been subsequently developed into versions for different regions and diverse ecosystems, distinguished by various measures and metric expectations within each category of measurement (Simon and Lyons 1995). Alternate versions for wadeable warm water streams in the central United States are being tested, and versions are being developed for coldwater streams, large unwadeable rivers, lakes, impoundments, and marine and Great Lakes estuaries.

Case Study

Watershed Name: Chasovenkov Verh Watershed
Location: Northern edge of Srednerusskaya Upland in the Tulu Region, Russia
Major Problem: Soil erosion and transport
Approaches: Sediment modelling to understand the fluvial dynamics.
Information:
http://www.fao.org/ag/aGL/watershed/watershed/papers/papercas/paperen/case9en.pdf

---

6.2.2 Biological Integrity

Biological integrity is defined as "the ability of an aquatic ecosystem to support and maintain a balanced, integrated, adaptive community of organisms having a species composition, diversity, and functional organization comparable to that of the natural habitats of a region" (Frey 1977).

Estimating biological integrity in surface waters based upon "least-impacted" conditions (unimpaired waters may no longer exist) has become the preferred method to guide restoration and protection programs. Biological criteria are defined as "numeric values or narrative expressions that describe the reference (least-impacted) biological integrity of aquatic communities inhabiting waters of a designated aquatic life use" (Davis and Simon 1995). These criteria are used to establish a multi-metric characterization of the aquatic community and as the basis for establishing restoration goals. The successful development and implementation of numeric biological criteria are based on: 1) developing a reference condition from a regional framework; 2) a multiple metric characterization of the aquatic community; and 3) a habitat evaluation.

The value of water resources is dependent not on chemical water quality and quantity alone, but also upon biological components (species) and the underlying biological processes that sustain those species. Because of this, a biological integrity/biocriteria assessment approach can be an effective means to manage for environmental results. Biological indicators — including species richness, species composition, population size, and trophic composition of resident organisms — can be used to effectively assess an aquatic ecosystem's biological integrity (Karr 1995).

Biological criteria Biological criteria provide guidelines for the sensitive tracking of resource conditions and are especially useful in assessing impairment of waters caused by nontoxic and nonchemical factors (which are predominant water pollutants). The U.S. EPA Science Advisory Board has agreed that biocriteria based on a multiple metric and regional reference approach can be used to 1) evaluate and demonstrate the success of current regulatory and management activities in protecting aquatic ecosystems; 2) provide a site-specific assessment of ecological degradation and ecosystem response to remediation or mitigation activities; and 3) assess biological resource trends in well-characterized watersheds (Davis and Simon 1995).

"The strengths of biocriteria-based biomonitoring include the ability to: assess and characterize resource status; diagnose and identify chemical, physical, and biological impacts as well as their cumulative effects; serve a broad range of environmental and regulatory programs when integrated with chemical and toxicity assessments; and provide a cost-effective approach to resource protection" (Karr 1995).

Multimetric approach The strength of the multimetric approach is in its ability to integrate information for the broad range of human impacts at the individual, population, community, and ecosystem levels. This approach also allows evaluation with reference to biogeography as a biologically based indicator of water resource quality. Multiple metric indices, such as the Index of Biotic Integrity (Karr 1981), can be used to accurately assess watershed health.

The broad outlines of the multimetric approach are as follows: Candidate matrices (based on a selected knowledge of aquatic systems), information regarding flora and fauna, literature reviews, and historical data, are selected and evaluated. The core metrics are chosen to represent diverse aspects of structure, composition, individual health, or processes of the aquatic biota. Collectively, these are the foundation for a sound, integrated analysis of the biotic condition (Barbour, Stribling, and Karr 1995). The fish Index of Biological Integrity (IBI) demonstrates the effectiveness of combining attributes to provide a valuable assessment of the status of aquatic ecosystems.

6.2.3 Habitat Index

Maintaining diverse, functional aquatic communities in surface waters requires preserving the natural physical habitat. Habitat, both instream and riparian, can be the limiting factor in determining the community potential of streams and rivers. The loss of habitat quality can result in extinctions, local extirpations, and population reductions of fish species and other aquatic flora and fauna (Rankin 1995). A habitat index which examines impacts from landscape ecosystems to microhabitat scales can be used to rank the physical and biological resource quality of streams, identify those threatened by human-induced change, and provide insight on possible remedies. Over the course of the last 20 to 30 years, habitat indices have been used to relate the standing crop or population of a target species (e.g. salmon) to habitat characteristics in a stream. Habitat indices help to determine the minimum or optimal streamflows that will protect habitat characteristics essential to the life history of one or more target species, and the indices can serve as a part of water pollution control programs.

Habitat indices that are based on a visual estimate of habitat features should include the following attributes:
- Substrate type and quality
- Instream physical structure/cover
- Channel structure/stability/modifications
- Riparian width/quality
- Bank erosion
- Flow/stream gradient
- Riffle-run/pool-glide quality/characteristics

6.2.4 Land Use Index

The land use index (LUI) is a planning method used to estimate the potential nonpoint source (NPS) pollutant contributions of surrounding land uses and landscape conditions to Wetland Evaluation Areas (WEA) (Carlisle 2002). A weighted index is used to gauge, estimate and rank the relative potential for cumulative impacts to affect different wetlands. Two key considerations factor into the use of the LUI: 1) the presence, state, and condition of three wetland zones of influence; namely, at 100 feet, 100 meters, and the wetland watershed (the area on the land that contributes surface and groundwater to the wetland); and 2) the assumption that the production, transport and fate of NPS pollutants are influenced by a number of determinants, including the nature and type of land use, the physical characteristics typical of certain land uses, hydrological patterns of the watershed, the contribution area of the WEA, and intercepting or attenuating conditions (Carlisle 2002).

Values of pollutant loading are estimated for each of the different land use types, or "sources" (recognizing that the actual pollutant loads generated from land use types will vary). The LUI method works as follows: 1) either GIS or a hand-held grid method is employed to examine the land use types in each zone of influence; 2) estimates of the extent of each land use type within the zones are derived and assigned land use loading coefficients; 3) an index score to indicate the relative potential for NPS production and transport to a given wetland is then calculated. The steps involved in deriving the LUI (Carlisle, 2002) are as follows:

(1) Preparation of a base map, delineating the WEA, the 100 foot (Zone 1) and 100 meter (Zone 2) zones, and the wetland watershed, and identification of the land uses.

(2) Completion of the Land Use Rapid Assessment Form, survey of the site and completion of the form, and derivation of the Rapid Assessment Form Score (summation of the question scores divided by 108 times 100 to arrive at the LUI score).

(3) Calculation of the Zone I and II areas using the LUI worksheet.

(4) Calculation of the extent of each different land use type in Zones I and II, using a worksheet and land use coefficient.

(5) Entering of the Final Zone scores and Rapid Assessment Form onto a worksheet, weighing each score and finding the average of the weighted scores (Carlisle 2002).

For more information on this method see Carlisle (2002).

6.3 WATER QUANTITY

Description

The population explosion in both developed and underdeveloped countries has created an enormous strain on the amount of fresh drinking water available in watersheds worldwide. Increased urbanization leads to the concentration of large populations in smaller areas of the watershed. These developed areas are rendered impervious to water infiltration, causing surface flows to be redirected away from dwindling groundwater sources. In order to accommodate urban water needs, storage reservoirs are created to redirect enormous amounts of surface flow.

Structural methods of evaluating aquifer recharge rates and surface flow patterns are being developed and used to find solutions to growing water needs.

Nonstructural methods aim at policies to increase available water, decrease demand through promoting more efficient household water use, and optimal water pricing.

Case Study

Watershed Name: Corrego do Paiol Watershed
Location: Sete Lagoas, Minas Gerais, Central-west Brazil
Major Problem: Water and soil conservation
Approaches: Micro-dams for rainfall water retention.
Information:
http://www.fao.org/ag/aGL/watershed/watershed/papers/papercas/paperen/case25en.pdf

---

6.3.1 Water Resource Assessment

A water resource assessment is an analysis of all of the potential ecosystem stresses and changes in water quality, as well as the social or economic impacts that will result from the proposed change in water use. Water use changes occur often when a new well is drilled; a new canal or irrigation ditch is dug, or when new structures are built that can affect waterway flow. Water resource assessments are designed to minimize poor water use, and other harmful environmental effects.

6.3.2 Hydrograph

The Unit Hydrograph is the most common method of evaluating and analyzing the runoff from storm events (Figure 6.2). The Unit Hydrograph (UHG) represents the watershed response to one unit of precipitation over a determined period of time and land area. The hydrograph is a plot of the volume of water (usually in m^3/sec), running through a stream versus time (usually hours) after the initial precipitation event. The inputs required for the formation of a UHG include precipitation quantity over time and fairly continuous monitoring of streamflow volume.

There are three main parts to a hydrograph: peak, rising limb, and receding limb. The peak is the highest point of the graph and represents the maximum rate of flow for a rainfall event. The curve before the peak is called the rising limb, which represents the increase in discharge from the time of initiation of rainfall at baseline river flow to the peak discharge volume. A short, steep rising limb means that the runoff from stormwater is not absorbed into the ground or slowed by vegetation. Therefore the peak occurs quickly and is quite high. The curve after the peak is the receding limb, which represents the decrease in discharge from peak volume back to base flow. From the receding limb, one can tell how fast the storm event has been assimilated and how long until normal flows return. During a storm event the Unit Hydrograph will rise steeply until the maximum flow is reached. As the watershed is able to absorb and pass the excess stormwater along, the curve will gradually decline until it returns to normal flow. In healthy watersheds, the hydrograph rises slowly and returns to base flow fairly rapidly.

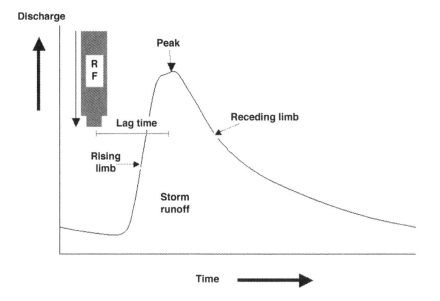

Figure 6.2 Example of a Unit Hydrograph

6.3.3 Habit Curve

A habit curve is a measure of water flow per unit of precipitation over a 24-hour period. It is used in urban areas to predict and plan sewer flow during storm events. The large amount of impervious surfaces in urban watersheds eliminates infiltration, increasing the volume of surface flow. Habit curves are used to assess the amount of underground sewer pipes needed to discharge water away from urban areas.

Case Study

Watershed Name: Tar-Pamlico River Watershed
Location: North Carolina, U.S.A.
Problem: Water quality
Approaches: Water quality trading between point and nonpoint sources of nutrients.
Information: http://www.epa.gov/owow/watershed/trading/tarpam.htm

---

6.3.4 Water Yield

Water is often the limiting resource for both human and animal populations. Most of the world's population receives its water from groundwater stored deep below the earth's surface. As populations increase, the demand for freshwater can become dangerously high. The safe water yield is the maximum amount of water that can be extracted from a watershed without exceeding the recharge rate for an extended period of time. Aquifers have different rates of recharge depending on soil types. Water can travel through sand and fine gravel at rates of 20-60 m/day, while in clay and dense rock water movement can be less than 0.001 m/day. Sustained depletion of groundwater can have many harmful effects on surrounding ecosystems.

Surface water and groundwater flow are intrinsically linked through the water cycle and must be evaluated in relation to one another. The presence of large volumes of water in healthy watersheds indicates that maximum infiltration has been reached and groundwater aquifers are at their highest capacities.

Low streamflow is an indication of a lowered water table. Extensive withdrawals from groundwater and lack of precipitation can lead to low groundwater levels. If the water table is lowered enough, it will no longer be possible for the groundwater to contribute to surface water flow in streams and rivers (Figure 6.3).

Ecosystem stress Decreasing levels of groundwater may affect the volume of surface streams, which can alter and destroy habitat in streambeds and riparian areas.

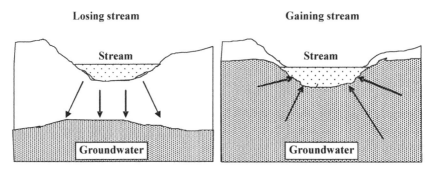

Figure 6.3 Flow contribution relationships of streams and groundwater

Flooding is an often-damaging level of streamflow that rises above stream banks and exceeds the channel capacity. Floodwaters may gather surface pollutants caught in above-ground catchments and disperse the pollution throughout the floodplains. However, flooding is necessary for the deposition of sediments throughout the floodplain, and for the replenishing of aquifers.

Groundwater recharge Management of the groundwater that is drawn up from wells is an area of extreme concern for densely populated regions. Groundwater is mostly recharged from precipitation infiltrating through the soil into the water table. During flood events the floodplains serve to enlarge the surface area exposed to water, greatly increasing the amount of water that is able to infiltrate underground aquifers. Influent streams feed aquifers through the pressure of gravity and the hydraulic gradient drawing water down to the groundwater. Changing the hydraulic pressure of groundwater can cause sinkholes and collapsing parts of the ground resulting in unpredictable environmental and economic damages.

SUGGESTED READING

6.1 – 6.2 Water Quality and Biological Monitoring

Barbour, Michael T., Stribling, James B. and Karr, James A. (1995) A Multimetric Approach for Establishing Biocriteria and Measuring Biological Condition. In *Biological Assessment and Criteria: Tools for Water Resource Planning and Decision Making*, (eds. Wayne A. Davis and Thomas P. Simon), pp. 63-77, Lewis Publishers, Boca Raton, Florida.

Carlisle. B.K. (2002) *Land Use Index: Wetland Evaluation Version - Description and Methodology.* Massachusetts Coastal Zone Management, Wetland Assessment Program, Boston, MA.
URL: http://www.salemsound.org/MA-CZM_Carlisle_LUI_1102.pdf

Davis, Wayne A. and Simon, Thomas P. (eds.) (1995) Introduction. In *Biological Assessment and Criteria: Tools for Water Resource Planning and Decision Making*, pp. 1-6, Lewis Publishers, Boca Raton, Florida.

DeShon, Jeffrey E. (1995) Development and Application of the Invertebrate Community Index. In *Biological Assessment and Criteria: Tools for Water Resource Planning and Decision Making*, (eds. Wayne A. Davis and Thomas P. Simon), pp. 217-243, Lewis Publishers, Boca Raton, Florida.

Frey, D. (1977) Biological integrity of water: an historical approach. In *The Integrity of Water*, (eds. R.K. Ballentine and L.J. Guarraia), pp. 127-140, Proceedings of a Symposium, March 10-12, 1975, U.S. Environmental Protection Agency, Washington, D.C.

Karr, J. R. (1981) Assessment of biotic integrity using fish communities. *Fisheries* **6**(6), 21-27.

Karr, James R. (1997) The Future Is Now: Biological Monitoring To Ensure Healthy Waters. *Northwest Science*, **71**(3), 254-257.

Karr, James R. (1995) Protecting Aquatic Ecosystems: Clean Water Is Not Enough. In *Biological Assessment and Criteria: Tools for Water Resource Planning and Decision Making*, (eds. Wayne A. Davis and Thomas P. Simon), pp. 7-13, Lewis Publishers, Boca Raton, Florida.

MacDonald, Lee H., Smart, Alan, W. and Wissmar, Robert C. (1991) *Monitoring Guidelines to Evaluate Effects of Forestry Activities on Streams in the Pacific Northwest and Alaska.* Report EPA 910/9-91-001, Region 10, U.S. Environmental Protection Agency, Seattle, Washington.

Meffe, Gary K. and Carroll, C. Ronald (1994) *Principles of Conservation Biology*, Sinauer Associates, Inc., Sunderland, Massachusetts.

Rankin, Edward T. (1995) Habitat Indices in Water Resource Quality Assessments. In *Biological Assessment and Criteria: Tools for Water Resource Planning and Decision Making*, (eds. Wayne A. Davis and Thomas P. Simon), pp. 181-208, Lewis Publishers, Boca Raton, Florida.

Simon, Thomas P. and Lyons, John (1995) Application of the Index of Biotic Integrity to Evaluate Water Resource Integrity in Freshwater Ecosystems. In *Biological Assessment and Criteria: Tools for Water Resource Planning and Decision Making,* (eds. Wayne A. Davis and Thomas P. Simon), pp. 245-262, Lewis Publishers, Boca Raton, Florida.

UNEP (2005b) *2005 - State of the UNEP GEMS/Water Global Network and Annual Report.* United Nations Environmental Program, Global Environmental Monitoring System – Water Program, Burlington, Ontario, Canada.

U.S. EPA (1996a) *Environmental Indicators of Water Quality in the United States.* Report EPA 841-R-96-002, U.S. Environmental Protection Agency, Washington, D.C.

U.S. EPA (1997b) *Volunteer Stream Monitoring: A Methods Manual.* Report EPA 841-B-97-003, U.S. Environmental Protection Agency, Office of Water (4503F), Washington, D.C.

6.3 Water Quantity

Annear, T., Chisholm, I., Beecher, H., Locke, A., Aarestad, P., Burkhart, N., Coomer, C., Estes, C., Hunt, J., Jacobson, R., Jobsis, G., Kauffman, Marshall, J., Mayes, K., Stalnaker, C. and Wentworth, R. (2002) *Instream Flows for Riverine Resource Stewardship*, Instream Flow Council, Cheyenne, Wyoming.

Burke, J.J. (2003) *Groundwater Management – The Search for Practical Approaches,* Water Reports 25, Food and Agriculture Organization of the United Nations, Rome. URL: ftp://ftp.fao.org/docrep/fao/005/y4502E/y4502E00.pdf.

Government of Alberta (2004) *Water Quantity Evaluation.* URL: http://www3.gov.ab.ca/env/water/gwsw/quantity/learn/evaluation/index.html.

Postel, S. and Richter, B. (2003) *Rivers for Life: Managing Water for People and Nature,* Island Press, Washington, D.C.

References

Adler, R.W., Landman, J.C. and Cameron, D.M. (1993) *The Clean Water Act 20 Years Later*, Island Press, Natural Resources Defense Council, Washington, D.C.

Agrawal, A. (2002) Common Resources and Institutional Sustainability. In *The Drama of the Commons*, (eds. E. Ostrom, T. Dietz, N. Dolsak, P. Stern, S. Stonich, and E. Weber), pp. 41-85, National Academy Press, Washington, D.C.

Agrios, George N. (1997) *Plant Pathology*, 4th edn, Academic Press, London, UK.

American Fisheries Society (1985) *Aquatic habitat Inventory: Glossary and Standards Methods*, (ed. W.T. Helm), pp. 1-34, Habitat Inventory Committee, Washington, D.C.

Annear, T., Chisholm, I., Beecher, H., Locke, A., Aarestad, P., Burkhart, N., Coomer, C., Estes, C., Hunt, J., Jacobson, R., Jobsis, G., Kauffman, J., Marshall, J., Mayes, K., Stalnaker, C. and Wentworth, R. (2002) *Instream Flows for Riverine Resource Stewardship*, Instream Flow Council, Cheyenne, Wyoming.

Baehr, A. and Corapcioglu, M. Yavuz (1984) A Predictive Model for Pollution from Gasoline in Soils and Groundwater. In *Petroleum Hydrocarbons and Organic Compounds in Ground water- Prevention Detection, and Restoration*, pp. 144-155, National Well Water Association, Worthington, Ohio.

Barbour, M.T., Gerritsen, J., Snyder, B.D. and Stribling, J.B. (1999) *Rapid Bioassessment Protocols for Use in Streams and Wadeable Rivers: Periphyton, Macroinvertebrates and Fish, Second Edition*. Report EPA 841-B-99-002, U.S. Environmental Protection Agency, Office of Water, Washington, D.C. URL: http://www.epa.gov/owow/monitoring/rbp/.

Barbour, Michael T., Stribling, James B. and Karr, James A. (1995) A Multimetric Approach for Establishing Biocriteria and Measuring Biological Condition. In *Biological Assessment and Criteria: Tools for Water Resource Planning and Decision Making*, (eds. Wayne A. Davis and Thomas P. Simon), pp. 63-77, Lewis Publishers, Boca Raton, Florida.

Barton, K. (1986) *Federal Wetlands Protection Programs*. Audubon Wildlife Report, National Audubon Society, New York.

Black, P. (1996) *Watershed Hydrology*, Ann Arbor Press, Chelsea, Michigan.

Brooks, Kenneth P., Ffolliott, H., Gregersen, H. and DeBano, L.F. (2003) *Hydrology and the Management of Watersheds,* 3rd edn, Iowa State Press, Ames, Iowa.

Brown K.W. and Donnelly, K.C. (1988) An estimation of risk associated with the organic constituents of hazardous and municipal waste landfill leachates. *Haz Waste Haz Matter.* **5**(1), 1-30.

Burke, J.J. (2003) *Groundwater Management – The Search for Practical Approaches*. Water Reports 25, Food and Agriculture Organization of the United Nations, Rome. URL: ftp://ftp.fao.org/docrep/fao/005/y4502E/y4502E00.pdf.

Carlisle, B.K. (2002) *Land Use Index: Wetland Evaluation Version - Description and Methodology*. Massachusetts Coastal Zone Management, Wetland Assessment Program, Boston, MA.
URL: http://www.salemsound.org/MA-CZM_Carlisle_LUI_1102.pdf.

Center for Watershed Protection (1998) *Rapid Watershed Planning Handbook: A Comprehensive Guide For Managing Urbanizing Watersheds*, Center for Watershed Protection, Inc., Elliott City, Maryland.

Center for Watershed Protection (2006a accessed) *Watershed Assessment*. URL: http://www.cwp.org/tools_assessment.htm.

Center for Watershed Protection (2006b accessed) *Watershed Restoration*. URL: http://www.cwp.org/restoration.htm.

Cowardin, L.M., Carter, V. F., Golet, C. and Laroe, E.T. (1979) *Classification of Wetlands and Deepwater Habitats of the United States*. Report FWS/OBS-79/32, U.S. Department of Interior, Fish and Wildlife Service, Washington, D.C.

Davis, Wayne A. and Simon, Thomas P. (eds.) (1995) Introduction. In *Biological Assessment and Criteria: Tools for Water Resource Planning and Decision Making*, pp. 1-6, Lewis Publishers, Boca Raton, Florida.

DeShon, Jeffrey E. (1995) Development and Application of the Invertebrate Community Index. In *Biological Assessment and Criteria: Tools for Water Resource Planning and Decision Making*, (eds. Wayne A. Davis and Thomas P. Simon), pp. 217-243, Lewis Publishers, Boca Raton, Florida.

Doppelt, B., Scurlock, M., Frissell, C. and Karr, J. (1993) *Entering the watershed, a new approach to save America's river ecosystems*, Island Press, Washington, D.C.

Etkin, D.S. (1997) *Oil Spills from Vessels (1960-1995): An International Historical Perspective,* Cutter Information Corporation, Cambridge, Massachusetts.

FISRWG (1998) *Stream Corridor Restoration: Principles, Processes and Practices*. Report GPO Item No. 0120-A; SuDocs No. A 57.6/2:EN3/PT.653. ISBN-0-934213-59-3, Federal Interagency Stream Restoration Working Group, Washington, D.C. URL: http://www.nrcs.usda.gov/technical/stream_restoration.

Forman, Richard T.T. (1995) *Land Mosaics: The ecology of landscapes and regions*, Cambridge University Press, West Nyack, New York.

Franklin, Hampden, Hampshire Conservation Districts (1997) *Massachusetts Erosion and Sediment Control Guidelines for Urban and Suburban Areas: A Guide for Planners, Designers and Municipal Officials*. Prepared for Massachusetts Department of Environmental Protection, Executive Office of Environmental Affairs, Northampton, Massachusetts.

Frey, D. (1977) Biological integrity of water: an historical approach. In *The Integrity of Water*, (eds. R.K. Ballentine and L.J. Guarraia), pp. 127-140, Proceedings of a Symposium, March 10-12, 1975, U.S. Environmental Protection Agency, Washington, D.C.

Government of Alberta (2004) *Water Quantity Evaluation*. URL: http://www3.gov.ab.ca/env/water/gwsw/quantity/learn/evaluation/index.html.

Houck, O.A. (2002) *Clean Water Act TMDL Program: Law, Policy, and Implementation*, 2nd edn, Environmental Law Institute, Washington, D.C.

Karr, J. R. (1981) Assessment of biotic integrity using fish communities. *Fisheries* **6**(6), 21-27.

Karr, James R. (1995) Protecting Aquatic Ecosystems: Clean Water Is Not Enough. In *Biological Assessment and Criteria: Tools for Water Resource Planning and Decision Making*, (eds. Wayne A. Davis and Thomas P. Simon), pp. 7-13, Lewis Publishers, Boca Raton, Florida.

Karr, James R. (1997) The Future Is Now: Biological Monitoring To Ensure Healthy Waters. *Northwest Science*, **71**(3), 254-257.

Kittredge, David B. Jr. and Parker, Michael (1999) *Massachusetts Forestry Best Management Practices Manual*. MA Department of Environmental Protection (MA DEP), Boston, Massachusetts.

LIH Landscape Information Hub (2006 accessed) *Landscape Planning Web: Links*. Univ. of Greenwich, London. URL:http://www.landscapeplanning.gre.ac.uk/links.htm.

London Biodiversity Partnership (2006) *Lakes, ponds and habitat audit*. URL: http://www.lbp.org.uk/02audit_pages/au13_ponds.html.

MA DEP (1993) *Landfill Technical Guidance Manual*. MA Department of Environmental Protection, Boston, Massachusetts.

MA DEP (2006) *Massachusetts Nonpoint Source Pollution Management Manual*. URL: http://projects.geosyntec.com/megamanual/.

MacDonald, Lee H., Smart, Alan, W. and Wissmar, Robert C. (1991) *Monitoring Guidelines to Evaluate Effects of Forestry Activities on Streams in the Pacific Northwest and Alaska*. Report EPA 910/9-91-001, Region 10, U.S. Environmental Protection Agency, Seattle, Washington.

Maryland Department of Natural Resources (2006 accessed) *Watershed Restoration Action Strategies*. URL: http://www.dnr.state.md.us/watersheds/wras/index.html.

Massachusetts Bays Program (1998) *Massachusetts Bays Watershed Stewardship Guide: An Education Resource*. URL: http://www.msp.umb.edu/mbea/mbeaguid.htm.

Massachusetts Bays Program (2003) *Massachusetts Bays Comprehensive Conservation & Management Program: An Evolving Plan For Action.*
URL: http://www.mass.gov/envir/massbays/pdf/revisedccmp.pdf.

Massachusetts Department of Environmental Protection (MA DEP) (2002) *Massachusetts Watershed Initiative Program*. MA Department of Environmental Protection, Worcester, MA.

Meffe, Gary K. and Carroll, C. Ronald (1994) *Principles of Conservation Biology*, Sinauer Associates, Inc., Sunderland, Massachusetts.

Miller, G. Tyler, Jr. (2006) *Environmental Science*, 11th edn, Wadsworth Publishing Co., Belmont, California.

Mitsch, William J. and Gosselink, James G. (2000) *Wetlands,* 3rd edn, John Wiley & Sons, New York.

National Aeronautics and Space Administration (NASA) (2004) *Global Hydrology Resource Center*. NASA, Global Hydrology and Climate Center. URL: http://ghrc.msfc.nasa.gov/.

National Research Council (1995) *Mexico City's Water Supply: Improving the Outlook for Sustainability*, Academia Nacional de Ingeniería, National Academy Press, Washington, D.C.

Natural Resources Conservation Service (NRCS) (2006 accessed) *How to Read a Topographic Map and Delineate a Watershed*. U.S. Department of Agriculture, NRCS. URL: http://www.nh.nrcs.usda.gov/technical/Publications/Topowatershed.pdf.

Oregon State University (2006 accessed) *Integrated Plant Protection Center*. URL: http://www.ipmnet.org/IPM_Handbooks.htm.

Ostendorf, D.W., Noss, R.R., and Lederer, D.O. (1984) *Landfill Leachate Migration through Shallow Unconfined Aquifers*. Water Resources Research. **20**, 291-296.

Ostrom, E. (1990) *Governing the Commons: The Evolution of Institutions for Collective Action*, Cambridge University Press, Cambridge, UK.

Postel, S. and Richter, B. (2003) *Rivers for Life: Managing Water for People and Nature*, Island Press, Washington, D.C.

Powell, J. W. (1890) Institutions for arid lands. *The Century*. **40**, 111–116.

Radcliffe, E.B. and Hutchison, W.D. (2006) (eds.), *Radcliffe's IPM World Textbook*, University of Minnesota, St. Paul, Minnesota. URL: http://ipmworld.umn.edu.

Randhir, T.O., and Genge, C. (2005) Watershed-based, Institutional Approach to Developing Clean Water Resources. *Journal of American Water Resources Association*. **41**(2), 413-424.

Rankin, Edward T. (1995) Habitat Indices in Water Resource Quality Assessments. In *Biological Assessment and Criteria: Tools for Water Resource Planning and Decision Making*, (eds. Wayne A. Davis and Thomas P. Simon), pp. 181-208, Lewis Publishers, Boca Raton, Florida.

Rothwell, R.L. (1983) Erosion and sediment production at road-stream crossings. *Forestry Chronicle*. **23**, 62-66.

Ryan, David K., Duggan, John W. and Bruell, Clifford J. (1995) *Enhanced Recovery of Gasoline Hydrocarbons by Soil Flushing with Solutions of Dissolved Organic Matter and Nonionic Surfactants*. Report 170, Water Resources Research Center, University of Massachusetts, Amherst, Massachusetts.

Schueler, Thomas R. and Holland, Heather K. (2000) The Importance of Imperviousness. In *The Practice of Watershed Protection,* (eds. Thomas R. Schueler and Heather K. Holland), pp. 7-16, Center for Watershed Protection, Ellicott City, Maryland.

Simon, Thomas P. and Lyons, John (1995) Application of the Index of Biotic Integrity to Evaluate Water Resource Integrity in Freshwater Ecosystems. In *Biological Assessment and Criteria: Tools for Water Resource Planning and Decision Making,* (eds. Wayne A. Davis and Thomas P. Simon), pp. 245-262, Lewis Publishers, Boca Raton, Florida.

Tiner, Ralph Jr. (1984) *Wetlands of the United States: Current status and recent trends*, U.S. Department of Interior, Fish and Wildlife Service, Washington, D.C.

U.S. Department of Agriculture (USDA) (1994) *Evaluating the Effectiveness of Forestry Best Management Practices in Meeting Water Quality Goals or Standards*. Report 1520, USDA, U.S. Forest Service, Southern Region, 1720 Peachtree Road NW, Atlanta, Georgia, U.S.A.

U.S. EPA (1991) *Volunteer Lake Monitoring*. Report EPA 440-4-91-002, U.S. Environmental Protection Agency, Office of Water, Washington DC, U.S.A. URL: http://www.epa.gov/volunteer/lake/lakevolman.pdf.

U.S. EPA (1993) *Guidance specifying management measures for sources of nonpoint pollution in coastal waters*. Report EPA 840-B-92-002, U.S. Environmental Protection Agency, Office of Water, Washinghton DC, U.S.A.

U.S. EPA (1994) *Developing Successful Runoff Control Programs For Urbanized Areas*. Report EPA 841-K-94-003, U.S. Environmental Protection Agency, Office of Water, Washington, D.C.

U.S. EPA (1996a) *Environmental Indicators of Water Quality in the United States*. Report EPA 841-R-96-002, U.S. Environmental Protection Agency, Washington DC, U.S.A.

U.S. EPA (1996b) *Municipal Wastewater Management Fact Sheets: Stormwater Best Management Practices*. Report EPA 832-F-96-001, U.S. Environmental Protection Agency, Washington DC, U.S.A.

U.S. EPA (1996c) *Watershed Progress: New York City Watershed Agreement.* Report EPA 840-F-96-005, U.S. Environmental Protection Agency, Office of Water, Washington, D.C.

U.S. EPA (1997a) *Compendium of Tools for Watershed Assessment and TMDL Development.* Report EPA 841-B-97-006, U.S. Environmental Protection Agency, Washington, D.C.

U.S. EPA (1997b) *Volunteer Stream Monitoring: A Methods Manual.* Report EPA 841-B-97-003, U.S. Environmental Protection Agency, Office of Water (4503F), Washington, D.C.

U.S. EPA (1998) *Clean Water Action Plan: Restoring and Protecting America's Waters.* Report EPA 840-R-98-001, U.S. Environmental Protection Agency, Office of Water, Washington, D.C.

U.S. EPA (1999a) *Protocol for Developing Nutrient TMDLs.* Report EPA 841-B-99-007, U.S. Environmental Protection Agency, Washington, D.C.

U.S. EPA (1999b) *Protocol for Developing Sediment TMDLs.* Report EPA 841-B-99-004, U.S. Environmental Protection Agency, Washington, D.C.

U.S. EPA (2000) *Low impact development: A literature review.* Report EPA 841B00005, Office of Water, U.S. Environmental Protection Agency, Washington, DC.

U.S. EPA (2000) *Stressor Identification Guidance Document.* Report EPA-822-B-00-025, U.S. Environmental Protection Agency, Office of Water, Washington, D.C. URL: http://www.epa.gov/waterscience/biocriteria/stressors/stressorid.pdf.

U.S. EPA (2001) *Protocol for Developing Pathogen TMDLs.* Report EPA 841-R-00-0002, U.S. Environmental Protection Agency, Washington, D.C.

U.S. EPA (2001) *The Brownfields Economic Redevelopment Initiative, Proposal Guidelines for Brownfields Cleanup.* Report EPA 500-F-01-348, Solid Waste and Emergency Response, Washington, D.C. URL: http://72.14.207.104/search?q=cache:_gYymWlOLxMJ:www.epa.gov/brownfields/pdf/bcrlfgui.pdf+Environmental+Protection+Agency.+1996.+The+Brownfield+Economic+Redevelopment+Initiative%3B+Application+Guidelines+for+Brownfields+Assessment+Demonstration+Pilots.&hl=en&gl=us&ct=clnk&cd=2&client=firefox-a.

U.S. EPA (2002) *Small Business Liability Relief and Brownfields Revitalization Act.* URL: http://www.epa.gov/swerosps/bf/sblrbra.htm.

U.S. EPA (2003) *Introduction to the Clean Water Act.* U.S. Environmental Protection Agency. URL: http://www.epa.gov/watertrain/cwa/.

U.S. EPA (2004) *2004-National Listing of Fish Advisories.* Fish Advisory Program, U.S. Environmental Protection Agency, Washington, D.C. URL: http://www.epa.gov/waterscience/fish/advisories/.

U.S. EPA (2005) *Draft Handbook for Developing Watershed Plans to Restore and Protect Our Waters.* Report EPA 841-B-05-005, U. S. Environmental Protection Agency, Office of Water, Washington, D.C.

U.S. EPA (2006a accessed) *Coastal Watershed Fact Sheets.* URL: http://www.epa.gov/owow/oceans/factsheets/fact5.html.

U.S. EPA (2006b accessed) *Estuaries and Near Coastal Areas, Bioassessment and Biocriteria Guidance.* URL: http://www.epa.gov/ost/biocriteria/States/estuaries/estuaries1.html.

U.S. EPA (2006c) *Examples of Approved TMDLs.* U.S. Environmental Protection Agency. URL: http://www.epa.gov/owow/tmdl/ examples/

U.S. EPA (2006d accessed) *How to Conserve Water and Use It Effectively.* U.S. Environmental Protection Agency, Office of Water. URL: http://www.epa.gov/ OW/you/chap3.html.

U.S. EPA (2006e accessed) *Lake and Reservoir Bioassessment and Biocriteria.* Technical Guidance Document. U.S. Environmental Protection Agency, Office of Water, Washington, D.C. URL: http://www.epa.gov/owow/ monitoring/tech/lakes.html.

U.S. EPA (2006f accessed) *Superfund, Cleaning Up the Nation's Hazardous Waste Sites.* URL: http://www.epa.gov/superfund/.

U.S. EPA (2006g accessed) *Surf Your Watershed.* U.S. Environmental Protection Agency. URL: http://www.epa.gov/surf/.

U.S. EPA (2006h) *Useful Links to Invasive Species Information.* URL: http://www.epa.gov/owow/invasive_species/links.html.

U.S. EPA (2006i accessed) *Watershed Academy.* U.S. Environmental Protection Agency. URL: http://www.epa.gov/watertrain/whywatersheds.html.

U.S. EPA (2006j) *Watershed Academy Web.* U.S. Environmental Protection Agency. URL: http://www.epa.gov/watertrain/watershedmgt/.

U.S. EPA (2006k accessed) *Watershed Approach Framework.* U.S. Environmental Protection Agency. URL: http://www.epa.gov/owow/watershed/framework.html.

U.S. EPA (2006l accessed) *Watershed Information Network.* U.S. Environmental Protection Agency. URL: http://www.epa.gov/win/.

U.S. Fish and Wildlife Service (1970) *National Estuary Study.* U.S. Department of Interior, Washington, D.C.

UNEP (2005a) *Inland Water Biodiversity.* Secretariat of the Convention on Biological Biodiversity, United Nations Environmental Program, UNEP-CBD, Montreal, Canada.

UNEP (2005b) *2005 - State of the UNEP GEMS / Water Global Network and Annual Report.* United Nations Environmental Program, Global Environmental Monitoring System – Water Program, Burlington, Ontario, Canada.

UNESCO (2006b) *Water a shared responsibility.* The United Nations World Water Development Report 2.
    URL: http://www.unesco.org/water/ wwap/wwdr2/table_contents.shtml.

UNESCO (2006c) *World Water Assessment Programme.* URL: http://www.unesco.org/water/wwap/description/index.shtml.

United Nations (UN) (1992) *Agenda 21: Chapter 18, Protection of the quality and supply of freshwater resources: Application of integrated approaches to the development, management and use of water resources.* UN Department of Economic and Social Affairs, Division for Sustainable Development. URL: http://www.un.org/ esa/sustdev/documents/agenda21/english/agenda21chapter18.htm.

United Nations Educational, Scientific and Cultural Organization (UNESCO) (2006a accessed) *Flow Regimes from International Experimental and Network Data (FRIEND).* International Hydrological Programme. URL: http://typo38.unesco.org/en/about-ihp/ihp-partners/assessment.html.

USDA (1995) *Wetland Values and Trends.* RCA Issue Brief #4, U.S. Depertment of Agriculture, Natural Resources Conservation Service, Washington, D.C.

USDA (1998a) *National Extension Targeted Water Quality Program, 1992-1995, Outcomes of Animal Waste Programs.* vol. 2, U.S. Department of Agriculture, Cooperative Extension Service, Washington, D.C.

USDA (1998b) *National Extension Targeted Water Quality Program, 1992-1995, Outcomes of Crop Pesticide Management Programs.* vol.4, U.S. Department of Agriculture, Cooperative Extension Service, Washington, D.C.

USDA (1998c) *National Extension Targeted Water Quality Program, 1992-1995, Outcomes of Nitrogen Fertilizer Management Programs.* vol.3, U.S. Department of Agriculture, Cooperative Extension Service, Washington, D.C.

Westbrooks, R. (1998) *Invasive plants, changing the landscape of America: Fact book.* The Federal Interagency Committee for the Management of Noxious and Exotic Weeds (FICMNEW), Washington, D.C.

Winkler, E.S. (1998) *Innovative and Alternative On-Site Wastewater Treatment Technologies Handbook.* UMASS Extension Bulletin Center, University of Massachusetts Amherst, Amherst, Massachusetts.

WCED (World Commission on Environment and Development) (1987) *Our Common Future,* Oxford University Press, Oxford, UK.

World Resources Institute (2003) *Watersheds of the World – Primary Watershed Map, Water Resources Atlas.* Water Resources Institute, Washington, D.C.

Worrall, Jim (2005) *Selected References in Forest Pathology.* Forest and Shade Tree Pathology. URL: http://www.forestpathology.org/refs.html

Glossary

Aquifer An underground water-bearing bed or permeable rock or earth material capable of holding and yielding large amounts of water.

Bankfull stage Stage at which a stream first overflows its natural bank.

Base flow Water flow coming from groundwater seepage into a stream.

Best Management Practices (BMPs) Techniques and field activities that provide the most effective means of reducing pollution.

Biodiversity The variety of life forms, especially the number of species, but also including the number of ecosystem types and genetic variation within species.

Bioremediation Use of living organisms in the process of cleaning up oil spills and contamination.

Desalination The removal of salts from saline water to convert it to freshwater.

Ecosystem A relatively homogenous area of organisms interacting with their environment.

Floodplain Riparian area that is covered by flood water during a storm event.

Grey water Wastewater generated from home use, except toilets. This includes waste from dishwashers, showers, sinks, and laundry.

Hydrograph A graphical representation of water discharge with respect to time.

Hydrologic cycle The cyclical transfer of water from the earth's surface into the atmosphere as evapotranspiration, from the atmosphere back to earth as precipitation, from surface to groundwater systems as infiltration, and from upstream to downstream as runoff into streams, rivers, and lakes, and ultimately into the oceans.

Hydrology The science of water, concerning the origin, circulation, distribution, and properties of the waters of the earth.

Landscape ecology A study of the structure, function, and change in a heterogeneous land area composed of interacting ecosystems.

Land subsidence Dropping of the land surface as a result of excessive pumping of groundwater.

Nonpoint source pollution Pollution that occurs over a wide area and that is usually associated with land use activities such as agricultural cultivation, grazing of livestock, urban runoff and forest management practices. These sources are variable, unpredictable, and dispersed as compared to point sources, which are steady, predictable, and concentrated.

Phreatic divide Subsurface divide that acts as a boundary for the flow of groundwater. Groundwater tends to move away in both directions that are normal to the ridge line. It is analogous to the topographical divide for surface flows in a watershed.

Riparian Pertaining to the bank of a stream or a water body.

River basin Similar to a watershed but larger in size. For example, the Mississippi River basin, the Amazon River basin and the Congo River basin comprise all the lands that drain through those rivers and their tributaries into the ocean.

Stream order A method of numbering streams in a drainage network. The smallest unbranched tributary is the first order stream, the stream receiving it is called the second order, and so on.

Sustainability The condition of maintaining ecological integrity and basic human needs over human generations.

Time of concentration The time required for water to flow from the farthest point in the watershed to a gauging station.

Watershed A topographically delineated area drained by a stream system; that is, the total land area above some point on a stream or river that drains past that point. The watershed is a hydrologic unit often used as a physical-biological unit and a socioeconomic-political unit for planning and management of natural resources.

Watershed management The process of organizing and guiding land and other resource use in a watershed to provide desired goods and services without adversely affecting soil and water resources. Embedded in the concept of watershed management is the recognition of the interrelationships among land use, soil, and water, and the links between uplands and downstream areas.

Watershed restoration Restoration is the reestablishment of the structure and function of an ecosystem. Structure refers to an ecosystem's native species diversity. Function refers to an ecosystem's productivity and the performance of its hydrology, trophic structure, and transport.

Wetland An area that is periodically inundated or saturated by surface or groundwater on an annual or seasonal basis, that displays hydric soils, and that typically supports or is capable of supporting hydrophytic vegetation.

Index

Lightning Source UK Ltd.
Milton Keynes UK
UKOW06n1857230815

257329UK00001B/47/P